# The Captain's Logbook
## *A Record of Astronomical Observation*

*By Josh Urban*

**ISBN 979-8-9864197-0-1**

1. Science/Nature
2. Astronomy-Handbooks, Manuals, etc

**Published by Telescopium Litteratura**

# The Captain's Logbook

## A Record of Astronomical Observation

### Josh Urban

*"Seeing is in some respect an art, which must be learnt."*
*– William Herschel*

*This work is dedicated to Jim Rosenstock, for instilling a love of visual observing, my father, Bob Urban, for showing me the constellations, and terri st. cloud, my high school English teacher.*

Special thanks to Darryl, Pam & Jeff, Frank, SMAS, John, Ardyce, Andrew, Bob, Chase, and Zakk.

An overdue tip of the hat to Noah Urban for the log sheet design, and excellent technical edits.
www.Mazuzu.com

*This Logbook belongs to:*

*Observations from:* _____

*To:* _____

*Personal Catalog Number:* _____

# Table of contents

**Introduction**

**Chapter 1:** How to Observe..................................................1

**Chapter 2:** The Records Dept. / Log Sheets........................3

**Chapter 3:** Details.............................................115

**Chapter 4:** Observing Projects.............................122

**Chapter 5:** Resources.......................................125

**Appendix:** Observing Checklists...........................127

# Introduction

**Ahoy, fellow stargazer!** The average person, warm and rested, might think us strange for standing in the cold at dark hours. But if I may wax poetic: We hear Infinity calling, and we cannot resist the summons. Captains of the sky, we pilot our "ships" (telescopes) outward to the Cosmos, sailing across Forever. Chilled, fatigued, and buoyed by elation, we collect ancient starlight from the Abyss. A dim red glow flickers across the charts as we plot a course to distant solar systems. Somewhere, an earthbound train whistles in the distance. Perhaps this is overly romantic, but *something* draws us outside on a clear night, transforming us into captains of our own stellar adventures.

Welcome to *The Captain's Logbook*. It's part journal, part workbook. It's written for anyone interested in improving their stargazing and documenting their journey across the Deep. There's also a few nifty observing projects. While aimed at the telescope or binocular user, all are welcome. I use a 12.5" reflector, 4" refractor, and 10x50 binoculars as aids, but my *eyes* are the primary instrument. The text is written for the Deep Sky, though the principles translate to other specialties. I suggest using this book in tandem with software or a good set of charts (I favor the *Uranometria 2000.0* and *Sky Atlas 2000.0.*)

Hopefully, the stars are calling, and you have light years to write about. Quit reading this, head outside and get to work! If you're like me, though, perhaps you like to make things complicated, so here's a bit about that.

# Chapter One: How to Observe

## Making It Complicated

To see more, you need a bigger telescope, right? "Doctor, Doctor, this 24" scope ain't cutting it. I think I have aperture fever. Is there any hope for me?" (Complications may include a strained back and empty bank account.)

Fortunately, I bring good news, friend! There's an active component to observing, proving the adage *hard work beats talent when talent doesn't work.* Skill matters in this game, and can be developed.

Firsthand experience has taught me this, and I'm here to pay it forward. As a competitive teenager, I'd do my best to keep up with the adults at the astronomy club. They showed me the tricks of the trade, such as using averted vision. Their patient guidance taught me to see in the dark. Summer nights would find me trying to outgun big scopes with my little 80mm refractor, and something happened: The closer I looked, the more I noticed.

I had started to learn guitar at about the same time, and was surprised that stargazing was a lot like music: practice makes progress. But *how* do we practice? *Writing detailed notes is a great starting place.* The next time you run across a new sight, ask yourself specific questions. *Exactly what does this object look like? Are there any hints of discernible structure? How does it sit in the star field? How would I describe this to a friend or a blind person?*

Have you noticed that a crowd of strangers looks different than a group of friends? *Anonymity blurs, specificity defines.* Taking detailed notes puts this phenomena to work for us. As an added bonus, we create a written record of our forward progress, which can inspire positive emotion. Moreover, even if you never read your notes, the act of defining and writing has been shown to be an effective way to "lock in" details to memory. But maybe that's a little heavy duty. In that case, I'd like to...

## Make it Fun

I'm a working DJ, and an avid fan of vinyl records, royal pain as they may be. They're big, heavy, fragile, prone to skip, and if you leave them in the car, they melt. Many's the time I've lugged my crate across a hot parking lot, sitting them inside the restaurant on the way back from a gig. But *man,* are they fun. There's the artwork, the feeling of a platter of music, the thunk of the needle dropping, the dust scratches...*the experience.*

This ethos is fun to take into the field. Notice your surroundings, and revel in them: the chill of the night air, and perhaps an owl hooting in the distance. Feel the cold metal of the scope bite your hand, and listen to the soft clunk as you change eyepieces on the night watch, gazing across Infinity. Hear the scratch of a pen as you jot notes like the astronomers of old, ancient photons raining down on the scene. Sure, you can log scientific catalog numbers, but don't forget to put down a few lines about the experience, your day, and what whimsical pattern the stars might make in the eyepiece. Adding a visceral element can heighten your powers of observation. Involve more than the visual, and have a blast while doing so. For one last angle, let's...

## Make it Philosophical

*God and Aliens. The Meaning of Life. How does our day relate to the Vast? What's out there?* Face time with the Abyss can inspire deep thoughts. Sprinkle them in next to the entries about atmospheric conditions and type of eyepiece you used to snag that new planetary nebula. This may be especially helpful on a frustrating night of observing. It's a shame to be grumpy in a field of cosmic diamonds (been there, done that!) The philosophical helps me keep things in perspective. A study of the Universe needn't be reduced to numbers on a checklist. But enough, already. Let's go outside, and catch some photons!

*Practice builds skill. The closer we look, the more we see.*

# Chapter Two: The Records Department

## Where to Start

You'll develop (or refine) your own personal style through practice. If you'd like some cues, try starting here, and modify as you wish. A quick shout-out to the beginners: no need to worry if you're unfamiliar with these terms. Any note is better than none at all. You'll also notice space for a sketch. Fine art degrees aren't needed - a simple jot goes a long way.

## Quick prompts: Discernment through Description

When at the eyepiece, ask:
- What does it look like?
- How bright is it (magnitude.)
- What details can I see? (Texture/patterns/hints of structure?)
- How "big" does it appear to be in the field of view?
- Are any colors detectable in the object or stars?
- How does the "neighborhood" (field of view) look?
- What are the atmospheric conditions? Is the sky steady and transparent?
- Is the Moon out? What's the phase?
- What's the weather like? Is it cold? Are there clouds?
- Are there any other interesting things happening on Earth? (Owls, frogs, etc.)
- What thoughts does it bring about? What feelings does it inspire?

## Here are the log sheets - time to get underway, sailor!

# Record
## of observation

**Date:**

**Time:**

| LOCATION | ............................................ |
| CONSTELLATION | ............................................ |
| OPTICS | ............................................ |

| OBJECT | ............................................ |
| FIRST SIGHTING | Y ☐ N ☐ ............................ |
| MAGNITUDE | ............................................ |

Seeing          Darkness

Transparency    Ambient Light

## Observational Notes:

Moon            Cold/Winter

At Home         Hot/Summer

At Observatory  Windy

On Road/Event   Calm

Company         Insects

Low Humidity    Clouds/Haze

High Humidity   Ground Fog

## Environmental Notes:

Eyepiece Sketch

## Personal Rating:

1   2   3   4   5   6   7   8   9   10

# Record
## of observation

**Date:**

**Time:**

| LOCATION | ............................................ |
| CONSTELLATION | ............................................ |
| OPTICS | ............................................ |

| OBJECT | ............................................ |
| FIRST SIGHTING | Y ☐ N ☐ ........................ |
| MAGNITUDE | ............................................ |

Seeing          Darkness

Transparency    Ambient Light

Moon            Cold/Winter

At Home         Hot/Summer

At Observatory  Windy

On Road/Event   Calm

Company         Insects

Low Humidity    Clouds/Haze

High Humidity   Ground Fog

## Observational Notes:

## Environmental Notes:

**Eyepiece Sketch**

## Personal Rating:

1    2    3    4    5    6    7    8    9    10

Josh Urban

# Record
## of observation

**Date:**

**Time:**

| | |
|---|---|
| **LOCATION** | |
| **CONSTELLATION** | |
| **OPTICS** | |

| | |
|---|---|
| **OBJECT** | |
| **FIRST SIGHTING** | Y ☐ N ☐ |
| **MAGNITUDE** | |

| | |
|---|---|
| Seeing | Darkness |
| Transparency | Ambient Light |
| Moon | Cold/Winter |
| At Home | Hot/Summer |
| At Observatory | Windy |
| On Road/Event | Calm |
| Company | Insects |
| Low Humidity | Clouds/Haze |
| High Humidity | Ground Fog |

## Observational Notes:

## Environmental Notes:

**Eyepiece Sketch**

## Personal Rating:

1   2   3   4   5   6   7   8   9   10

# Record
## of observation

**Date:**

**Time:**

| LOCATION | .................................................... |
| CONSTELLATION | .................................................... |
| OPTICS | .................................................... |

| OBJECT | .................................................... |
| FIRST SIGHTING | Y ☐ N ☐ .................................... |
| MAGNITUDE | .................................................... |

| Seeing | Darkness |
| Transparency | Ambient Light |
| Moon | Cold/Winter |
| At Home | Hot/Summer |
| At Observatory | Windy |
| On Road/Event | Calm |
| Company | Insects |
| Low Humidity | Clouds/Haze |
| High Humidity | Ground Fog |

## Observational Notes:

## Environmental Notes:

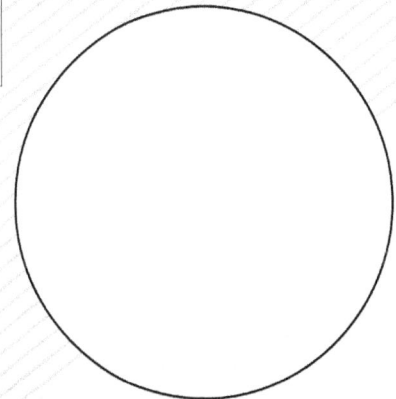

**Eyepiece Sketch**

## Personal Rating:

1    2    3    4    5    6    7    8    9    10

# Record
## of observation

**Date:**

**Time:**

| LOCATION | |
| CONSTELLATION | |
| OPTICS | |

| OBJECT | |
| FIRST SIGHTING | Y ☐ N ☐ |
| MAGNITUDE | |

Seeing      Darkness

Transparency      Ambient Light

Moon      Cold/Winter

At Home      Hot/Summer

At Observatory      Windy

On Road/Event      Calm

Company      Insects

Low Humidity      Clouds/Haze

High Humidity      Ground Fog

## Observational Notes:

## Environmental Notes:

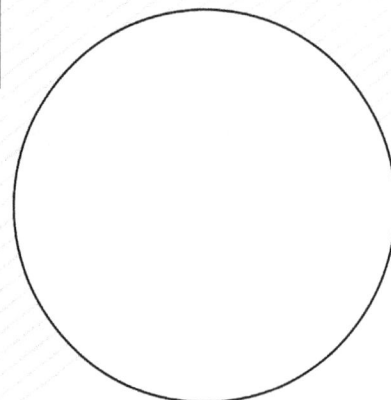

**Eyepiece Sketch**

## Personal Rating:

**1   2   3   4   5   6   7   8   9   10**

# Record
## of observation

**Date:**

**Time:**

| | |
|---|---|
| **LOCATION** | ........................................... |
| **CONSTELLATION** | ........................................... |
| **OPTICS** | ........................................... |

| | |
|---|---|
| **OBJECT** | ........................................... |
| **FIRST SIGHTING** | Y ☐ N ☐ ........................... |
| **MAGNITUDE** | ........................................... |

Seeing      Darkness

Transparency      Ambient Light

Moon      Cold/Winter

At Home      Hot/Summer

At Observatory      Windy

On Road/Event      Calm

Company      Insects

Low Humidity      Clouds/Haze

High Humidity      Ground Fog

## Observational Notes:

## Environmental Notes:

Eyepiece Sketch

## Personal Rating:

1    2    3    4    5    6    7    8    9    10

Josh Urban

# Record
## of observation

Date:

Time:

| LOCATION | .................................. |
| CONSTELLATION | .................................. |
| OPTICS | .................................. |

| OBJECT | .................................. |
| FIRST SIGHTING | Y ☐ N ☐ |
| MAGNITUDE | .................................. |

Seeing            Darkness

Transparency      Ambient Light

## Observational Notes:

Moon              Cold/Winter

At Home           Hot/Summer

At Observatory    Windy

On Road/Event     Calm

Company           Insects

Low Humidity      Clouds/Haze

High Humidity     Ground Fog

## Environmental Notes:

Eyepiece Sketch

## Personal Rating:

1    2    3    4    5    6    7    8    9    10

# Record
## of observation

**Date:**

**Time:**

| LOCATION | .................................... |
| CONSTELLATION | .................................... |
| OPTICS | .................................... |

| OBJECT | .................................... |
| FIRST SIGHTING | Y ☐ N ☐ |
| MAGNITUDE | .................................... |

Seeing    Darkness

Transparency    Ambient Light

Moon    Cold/Winter

At Home    Hot/Summer

At Observatory    Windy

On Road/Event    Calm

Company    Insects

Low Humidity    Clouds/Haze

High Humidity    Ground Fog

## Observational Notes:

## Environmental Notes:

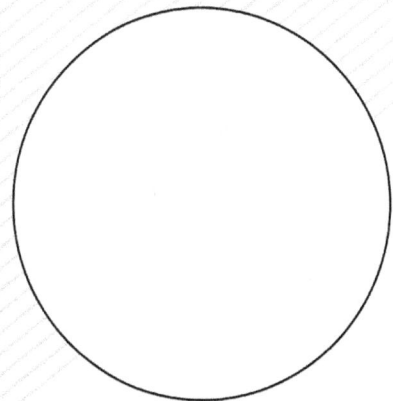

**Eyepiece Sketch**

**Personal Rating:**

1   2   3   4   5   6   7   8   9   10

# Record
## of observation

**Date:**

**Time:**

| LOCATION | ................................................. |
| CONSTELLATION | ................................................. |
| OPTICS | ................................................. |

| OBJECT | ................................................. |
| FIRST SIGHTING | Y ☐ N ☐ |
| MAGNITUDE | ................................................. |

## Observational Notes:

Seeing        Darkness

Transparency    Ambient Light

Moon        Cold/Winter

At Home     Hot/Summer

At Observatory  Windy

On Road/Event  Calm

Company    Insects

Low Humidity    Clouds/Haze

High Humidity   Ground Fog

## Environmental Notes:

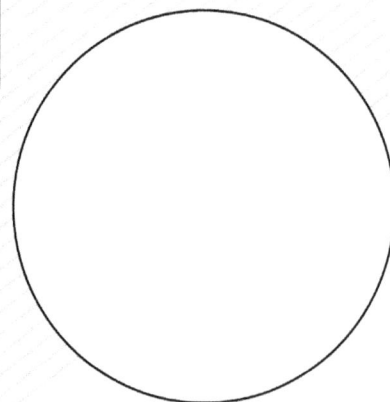

Eyepiece Sketch

## Personal Rating:

**1  2  3  4  5  6  7  8  9  10**

# Record
## of observation

**Date:**

**Time:**

| LOCATION | ............................................... |
| CONSTELLATION | ............................................... |
| OPTICS | ............................................... |

| OBJECT | ............................................... |
| FIRST SIGHTING | Y ☐ N ☐ ............................... |
| MAGNITUDE | ............................................... |

## Observational Notes:

Seeing      Darkness

Transparency      Ambient Light

Moon      Cold/Winter

At Home      Hot/Summer

At Observatory      Windy

On Road/Event      Calm

Company      Insects

Low Humidity      Clouds/Haze

High Humidity      Ground Fog

## Environmental Notes:

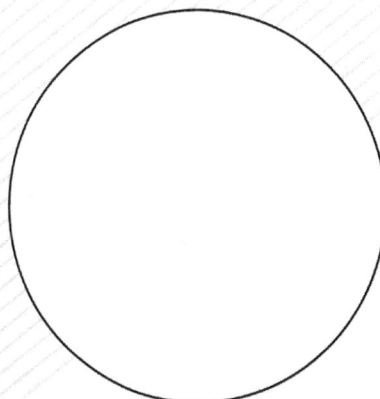

Eyepiece Sketch

## Personal Rating:

1    2    3    4    5    6    7    8    9    10

# Record
## of observation

**Date:**

**Time:**

| LOCATION | .................................... |
| CONSTELLATION | .................................... |
| OPTICS | .................................... |

| OBJECT | .................................... |
| FIRST SIGHTING | Y ☐ N ☐ |
| MAGNITUDE | .................................... |

| Seeing | Darkness |
|---|---|
| Transparency | Ambient Light |
| Moon | Cold/Winter |
| At Home | Hot/Summer |
| At Observatory | Windy |
| On Road/Event | Calm |
| Company | Insects |
| Low Humidity | Clouds/Haze |
| High Humidity | Ground Fog |

## Observational Notes:

## Environmental Notes:

Eyepiece Sketch

## Personal Rating:

1    2    3    4    5    6    7    8    9    10

# Record
## of observation

**Date:**

**Time:**

| LOCATION | ........................... |
| CONSTELLATION | ........................... |
| OPTICS | ........................... |

| OBJECT | ........................... |
| FIRST SIGHTING | Y ☐ N ☐ ........................... |
| MAGNITUDE | ........................... |

## Observational Notes:

| | |
|---|---|
| Seeing | Darkness |
| Transparency | Ambient Light |
| Moon | Cold/Winter |
| At Home | Hot/Summer |
| At Observatory | Windy |
| On Road/Event | Calm |
| Company | Insects |
| Low Humidity | Clouds/Haze |
| High Humidity | Ground Fog |

## Environmental Notes:

**Eyepiece Sketch**

## Personal Rating:

1    2    3    4    5    6    7    8    9    10

Josh Urban

# **Record**
## of observation

Date:

Time:

| LOCATION | |
|---|---|
| CONSTELLATION | |
| OPTICS | |

| OBJECT | |
|---|---|
| FIRST SIGHTING | Y ☐ N ☐ |
| MAGNITUDE | |

## Observational Notes:

| Seeing | Darkness |
|---|---|
| Transparency | Ambient Light |
| Moon | Cold/Winter |
| At Home | Hot/Summer |
| At Observatory | Windy |
| On Road/Event | Calm |
| Company | Insects |
| Low Humidity | Clouds/Haze |
| High Humidity | Ground Fog |

## Environmental Notes:

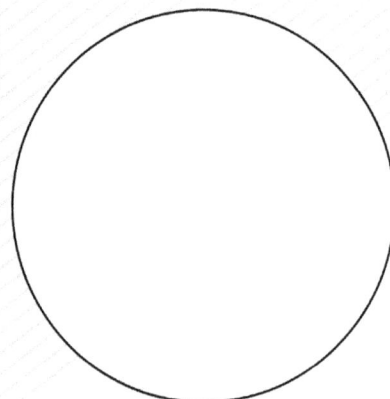

**Eyepiece Sketch**

## Personal Rating:

1    2    3    4    5    6    7    8    9    10

# Record
## of observation

**Date:**

**Time:**

| LOCATION | ........................ |
| CONSTELLATION | ........................ |
| OPTICS | ........................ |

| OBJECT | ........................ |
| FIRST SIGHTING | Y ☐ N ☐ |
| MAGNITUDE | ........................ |

## Observational Notes:

Seeing      Darkness

Transparency      Ambient Light

Moon      Cold/Winter

At Home      Hot/Summer

At Observatory      Windy

On Road/Event      Calm

Company      Insects

Low Humidity      Clouds/Haze

High Humidity      Ground Fog

## Environmental Notes:

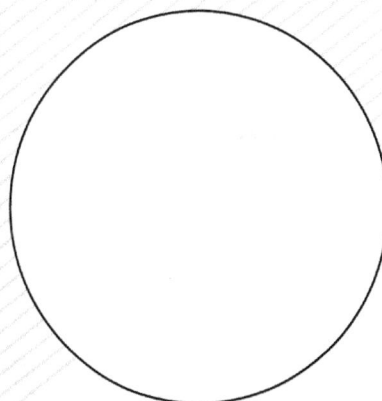

Eyepiece Sketch

## Personal Rating:

1   2   3   4   5   6   7   8   9   10

Josh Urban

# Record
## of observation

**Date:**

**Time:**

| LOCATION | ............................... |
| CONSTELLATION | ............................... |
| OPTICS | ............................... |

| OBJECT | ............................... |
| FIRST SIGHTING | Y ☐ N ☐ |
| MAGNITUDE | ............................... |

## Observational Notes:

| Seeing | Darkness |
| Transparency | Ambient Light |
| Moon | Cold/Winter |
| At Home | Hot/Summer |
| At Observatory | Windy |
| On Road/Event | Calm |
| Company | Insects |
| Low Humidity | Clouds/Haze |
| High Humidity | Ground Fog |

## Environmental Notes:

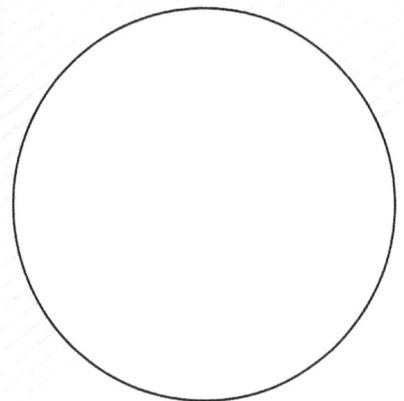

Eyepiece Sketch

## Personal Rating:

1    2    3    4    5    6    7    8    9    10

# Record
## of observation

**Date:**

**Time:**

| LOCATION | .................................... |
| CONSTELLATION | .................................... |
| OPTICS | .................................... |

| OBJECT | .................................... |
| FIRST SIGHTING | Y ☐ N ☐ |
| MAGNITUDE | .................................... |

Seeing          Darkness

Transparency    Ambient Light

Moon            Cold/Winter

At Home         Hot/Summer

At Observatory  Windy

On Road/Event   Calm

Company         Insects

Low Humidity    Clouds/Haze

High Humidity   Ground Fog

## Observational Notes:

## Environmental Notes:

**Eyepiece Sketch**

## Personal Rating:

1    2    3    4    5    6    7    8    9    10

# Record
## of observation

**Date:**

**Time:**

| | |
|---|---|
| **LOCATION** | ................................................. |
| **CONSTELLATION** | ................................................. |
| **OPTICS** | ................................................. |

| | |
|---|---|
| **OBJECT** | ................................................. |
| **FIRST SIGHTING** | Y ☐ N ☐ |
| **MAGNITUDE** | ................................................. |

Seeing          Darkness

Transparency          Ambient Light

## Observational Notes:

Moon          Cold/Winter

At Home          Hot/Summer

At Observatory          Windy

On Road/Event          Calm

Company          Insects

Low Humidity          Clouds/Haze

High Humidity          Ground Fog

## Environmental Notes:

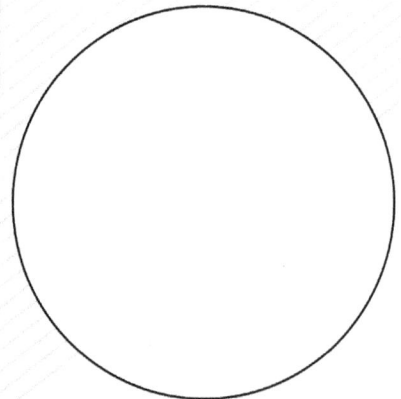

Eyepiece Sketch

## Personal Rating:

1    2    3    4    5    6    7    8    9    10

# Record
## of observation

**Date:**

**Time:**

| LOCATION | ............................................ |
| CONSTELLATION | ............................................ |
| OPTICS | ............................................ |

| OBJECT | ............................................ |
| FIRST SIGHTING | Y ☐ N ☐ |
| MAGNITUDE | ............................................ |

| Seeing | Darkness |
| Transparency | Ambient Light |
| Moon | Cold/Winter |
| At Home | Hot/Summer |
| At Observatory | Windy |
| On Road/Event | Calm |
| Company | Insects |
| Low Humidity | Clouds/Haze |
| High Humidity | Ground Fog |

## Observational Notes:

## Environmental Notes:

**Eyepiece Sketch**

## Personal Rating:

1    2    3    4    5    6    7    8    9    10

# Record
## of observation

**Date:**

**Time:**

| LOCATION | ............................................ |
| CONSTELLATION | ............................................ |
| OPTICS | ............................................ |

| OBJECT | ............................................ |
| FIRST SIGHTING | Y ☐ N ☐ ............................. |
| MAGNITUDE | ............................................ |

## Observational Notes:

| Seeing | Darkness |
| Transparency | Ambient Light |
| Moon | Cold/Winter |
| At Home | Hot/Summer |
| At Observatory | Windy |
| On Road/Event | Calm |
| Company | Insects |
| Low Humidity | Clouds/Haze |
| High Humidity | Ground Fog |

## Environmental Notes:

**Eyepiece Sketch**

## Personal Rating:

1    2    3    4    5    6    7    8    9    10

# Record
## of observation

**Date:**

**Time:**

| LOCATION | .................................................... |
| CONSTELLATION | .................................................... |
| OPTICS | .................................................... |

| OBJECT | .................................................... |
| FIRST SIGHTING | Y ☐ N ☐ |
| MAGNITUDE | .................................................... |

## Observational Notes:

| Seeing | Darkness |
| Transparency | Ambient Light |
| Moon | Cold/Winter |
| At Home | Hot/Summer |
| At Observatory | Windy |
| On Road/Event | Calm |
| Company | Insects |
| Low Humidity | Clouds/Haze |
| High Humidity | Ground Fog |

## Environmental Notes:

**Eyepiece Sketch**

## Personal Rating:

1    2    3    4    5    6    7    8    9    10

Josh Urban

# Record
## of observation

**Date:**

**Time:**

| | |
|---|---|
| **LOCATION** | ............................... |
| **CONSTELLATION** | ............................... |
| **OPTICS** | ............................... |

| | |
|---|---|
| **OBJECT** | ............................... |
| **FIRST SIGHTING** | Y ☐ N ☐ |
| **MAGNITUDE** | ............................... |

Seeing     Darkness

Transparency     Ambient Light

## Observational Notes:

Moon     Cold/Winter

At Home     Hot/Summer

At Observatory     Windy

On Road/Event     Calm

Company     Insects

Low Humidity     Clouds/Haze

High Humidity     Ground Fog

## Environmental Notes:

**Eyepiece Sketch**

## Personal Rating:

1    2    3    4    5    6    7    8    9    10

# Record
## of observation

**Date:**

**Time:**

| LOCATION | ..................................... |
| CONSTELLATION | ..................................... |
| OPTICS | ..................................... |

| OBJECT | ..................................... |
| FIRST SIGHTING | Y ☐ N ☐ |
| MAGNITUDE | ..................................... |

Seeing     Darkness

Transparency     Ambient Light

## Observational Notes:

Moon     Cold/Winter

At Home     Hot/Summer

At Observatory     Windy

On Road/Event     Calm

Company     Insects

Low Humidity     Clouds/Haze

High Humidity     Ground Fog

## Environmental Notes:

Eyepiece Sketch

## Personal Rating:

1    2    3    4    5    6    7    8    9    10

# Record
## of observation

**Date:**

**Time:**

| LOCATION | ............................................ | OBJECT | ............................................ |
| CONSTELLATION | ............................................ | FIRST SIGHTING | Y ☐ N ☐ ............................ |
| OPTICS | ............................................ | MAGNITUDE | ............................................ |

Seeing      Darkness

## Observational Notes:

Transparency      Ambient Light

Moon      Cold/Winter

At Home      Hot/Summer

At Observatory      Windy

On Road/Event      Calm

Company      Insects

Low Humidity      Clouds/Haze

High Humidity      Ground Fog

## Environmental Notes:

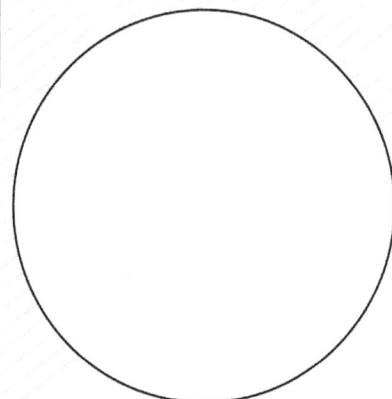

**Eyepiece Sketch**

## Personal Rating:

1    2    3    4    5    6    7    8    9    10

# Record
## of observation

**Date:**

**Time:**

| LOCATION | ........................................... |
| CONSTELLATION | ........................................... |
| OPTICS | ........................................... |

| OBJECT | ........................................... |
| FIRST SIGHTING | Y ☐ N ☐ |
| MAGNITUDE | ........................................... |

| Seeing | Darkness |
| Transparency | Ambient Light |
| Moon | Cold/Winter |
| At Home | Hot/Summer |
| At Observatory | Windy |
| On Road/Event | Calm |
| Company | Insects |
| Low Humidity | Clouds/Haze |
| High Humidity | Ground Fog |

## Observational Notes:

## Environmental Notes:

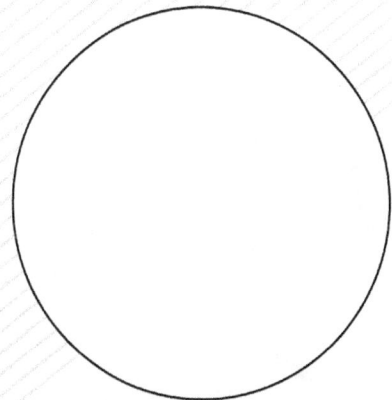

**Eyepiece Sketch**

## Personal Rating:

**1    2    3    4    5    6    7    8    9    10**

Josh Urban

# Record
## of observation

**Date:**

**Time:**

| LOCATION | ............................................... |
| CONSTELLATION | ............................................... |
| OPTICS | ............................................... |

| OBJECT | ............................................... |
| FIRST SIGHTING | Y ☐ N ☐ |
| MAGNITUDE | ............................................... |

Seeing     Darkness

Transparency     Ambient Light

Moon     Cold/Winter

At Home     Hot/Summer

At Observatory     Windy

On Road/Event     Calm

Company     Insects

Low Humidity     Clouds/Haze

High Humidity     Ground Fog

## Observational Notes:

## Environmental Notes:

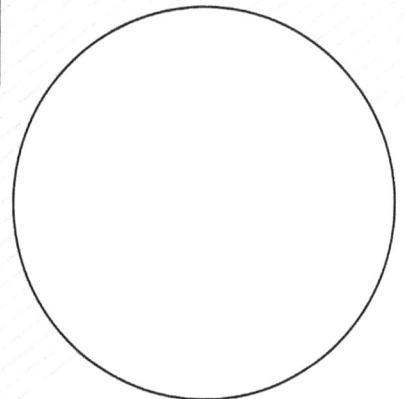

**Eyepiece Sketch**

## Personal Rating:

1    2    3    4    5    6    7    8    9    10

# Record
## of observation

**Date:**

**Time:**

| LOCATION | ............................................ |
| CONSTELLATION | ............................................ |
| OPTICS | ............................................ |

| OBJECT | ............................................ |
| FIRST SIGHTING | Y ☐ N ☐ |
| MAGNITUDE | ............................................ |

## Observational Notes:

Seeing            Darkness

Transparency      Ambient Light

Moon              Cold/Winter

At Home           Hot/Summer

At Observatory    Windy

On Road/Event     Calm

Company           Insects

Low Humidity      Clouds/Haze

High Humidity     Ground Fog

## Environmental Notes:

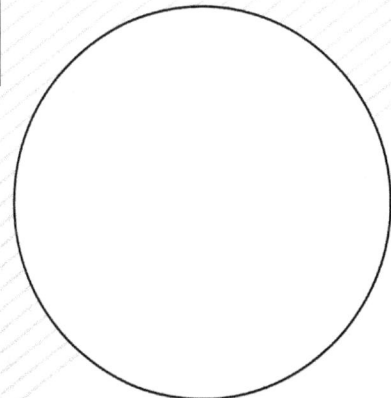

Eyepiece Sketch

## Personal Rating:

1    2    3    4    5    6    7    8    9    10

# Record
## of observation

**Date:**

**Time:**

| LOCATION | |
|---|---|
| CONSTELLATION | |
| OPTICS | |

| OBJECT | |
|---|---|
| FIRST SIGHTING | Y ☐ N ☐ |
| MAGNITUDE | |

| | |
|---|---|
| Seeing | Darkness |
| Transparency | Ambient Light |
| Moon | Cold/Winter |
| At Home | Hot/Summer |
| At Observatory | Windy |
| On Road/Event | Calm |
| Company | Insects |
| Low Humidity | Clouds/Haze |
| High Humidity | Ground Fog |

## Observational Notes:

## Environmental Notes:

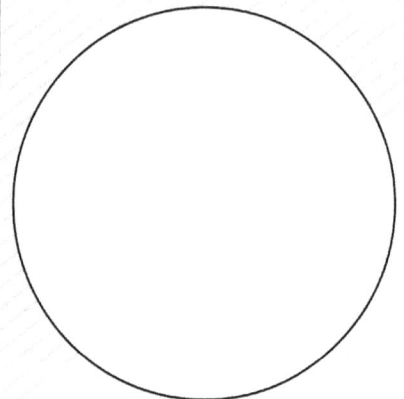

Eyepiece Sketch

## Personal Rating:

1    2    3    4    5    6    7    8    9    10

# Record
## of observation

**Date:**

**Time:**

| LOCATION | .................................... | OBJECT | .................................... |
| CONSTELLATION | .................................... | FIRST SIGHTING | Y ☐ N ☐ |
| OPTICS | .................................... | MAGNITUDE | .................................... |

## Observational Notes:

| Seeing | Darkness |
| Transparency | Ambient Light |
| Moon | Cold/Winter |
| At Home | Hot/Summer |
| At Observatory | Windy |
| On Road/Event | Calm |
| Company | Insects |
| Low Humidity | Clouds/Haze |
| High Humidity | Ground Fog |

## Environmental Notes:

**Eyepiece Sketch**

## Personal Rating:

**1    2    3    4    5    6    7    8    9    10**

Josh Urban

# Record
## of observation

**Date:**

**Time:**

| LOCATION | .............................. |
| CONSTELLATION | .............................. |
| OPTICS | .............................. |

| OBJECT | .............................. |
| FIRST SIGHTING | Y ☐ N ☐ |
| MAGNITUDE | .............................. |

## Observational Notes:

Seeing      Darkness

Transparency      Ambient Light

Moon      Cold/Winter

At Home      Hot/Summer

At Observatory      Windy

On Road/Event      Calm

Company      Insects

Low Humidity      Clouds/Haze

High Humidity      Ground Fog

## Environmental Notes:

Eyepiece Sketch

## Personal Rating:

1    2    3    4    5    6    7    8    9    10

# Record
## of observation

**Date:**

**Time:**

| LOCATION | |
|---|---|
| CONSTELLATION | |
| OPTICS | |

| OBJECT | |
|---|---|
| FIRST SIGHTING | Y ☐ N ☐ |
| MAGNITUDE | |

## Observational Notes:

Seeing          Darkness

Transparency    Ambient Light

Moon            Cold/Winter

At Home         Hot/Summer

At Observatory  Windy

On Road/Event   Calm

Company         Insects

Low Humidity    Clouds/Haze

High Humidity   Ground Fog

## Environmental Notes:

**Eyepiece Sketch**

## Personal Rating:

1    2    3    4    5    6    7    8    9    10

# Record
## of observation

**Date:**

**Time:**

| LOCATION | ................................................ |
| CONSTELLATION | ................................................ |
| OPTICS | ................................................ |

| OBJECT | ................................................ |
| FIRST SIGHTING | Y ☐ N ☐ |
| MAGNITUDE | ................................................ |

## Observational Notes:

Seeing      Darkness

Transparency      Ambient Light

Moon      Cold/Winter

At Home      Hot/Summer

At Observatory      Windy

On Road/Event      Calm

Company      Insects

Low Humidity      Clouds/Haze

High Humidity      Ground Fog

## Environmental Notes:

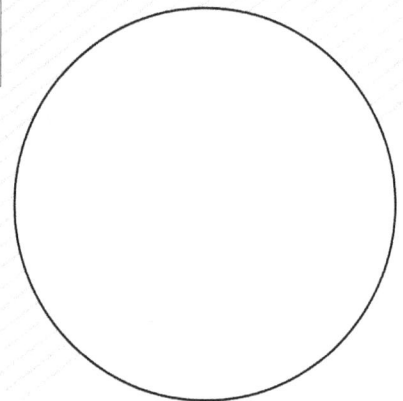

**Eyepiece Sketch**

## Personal Rating:

1    2    3    4    5    6    7    8    9    10

# Record
## of observation

**Date:**

**Time:**

| LOCATION | ............................................ |
| CONSTELLATION | ............................................ |
| OPTICS | ............................................ |

| OBJECT | ............................................ |
| FIRST SIGHTING | Y ☐ N ☐ ............................ |
| MAGNITUDE | ............................................ |

| | |
|---|---|
| Seeing | Darkness |
| Transparency | Ambient Light |
| Moon | Cold/Winter |
| At Home | Hot/Summer |
| At Observatory | Windy |
| On Road/Event | Calm |
| Company | Insects |
| Low Humidity | Clouds/Haze |
| High Humidity | Ground Fog |

## Observational Notes:

## Environmental Notes:

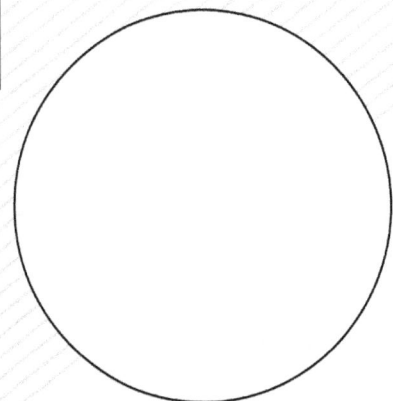

**Eyepiece Sketch**

## Personal Rating:

1  2  3  4  5  6  7  8  9  10

# Record
## of observation

**Date:**

**Time:**

| LOCATION | ............................ |
| CONSTELLATION | ............................ |
| OPTICS | ............................ |

| OBJECT | ............................ |
| FIRST SIGHTING | Y ☐ N ☐ ............................ |
| MAGNITUDE | ............................ |

## Observational Notes:

| | |
|---|---|
| Seeing | Darkness |
| Transparency | Ambient Light |
| Moon | Cold/Winter |
| At Home | Hot/Summer |
| At Observatory | Windy |
| On Road/Event | Calm |
| Company | Insects |
| Low Humidity | Clouds/Haze |
| High Humidity | Ground Fog |

## Environmental Notes:

Eyepiece Sketch

## Personal Rating:

**1    2    3    4    5    6    7    8    9    10**

# Record
## of observation

**Date:**

**Time:**

| LOCATION | .................................................... |
| CONSTELLATION | .................................................... |
| OPTICS | .................................................... |

| OBJECT | .................................................... |
| FIRST SIGHTING | Y ☐ N ☐ |
| MAGNITUDE | .................................................... |

| Seeing | Darkness |
| Transparency | Ambient Light |
| Moon | Cold/Winter |
| At Home | Hot/Summer |
| At Observatory | Windy |
| On Road/Event | Calm |
| Company | Insects |
| Low Humidity | Clouds/Haze |
| High Humidity | Ground Fog |

## Observational Notes:

## Environmental Notes:

Eyepiece Sketch

## Personal Rating:

1    2    3    4    5    6    7    8    9    10

Josh Urban

# **Record**
## of observation

**Date:**

**Time:**

| LOCATION | ............................... |
| CONSTELLATION | ............................... |
| OPTICS | ............................... |

| OBJECT | ............................... |
| FIRST SIGHTING | Y ☐ N ☐ |
| MAGNITUDE | ............................... |

Seeing      Darkness

Transparency      Ambient Light

Moon      Cold/Winter

At Home      Hot/Summer

At Observatory      Windy

On Road/Event      Calm

Company      Insects

Low Humidity      Clouds/Haze

High Humidity      Ground Fog

## Observational Notes:

## Environmental Notes:

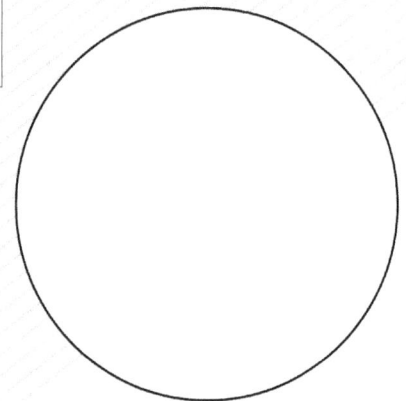

**Eyepiece Sketch**

## Personal Rating:

1    2    3    4    5    6    7    8    9    10

# Record
## of observation

**Date:**

**Time:**

| LOCATION | ............................................. |
| CONSTELLATION | ............................................. |
| OPTICS | ............................................. |

| OBJECT | ............................................. |
| FIRST SIGHTING | Y ☐ N ☐ |
| MAGNITUDE | ............................................. |

## Observational Notes:

Seeing            Darkness

Transparency      Ambient Light

Moon              Cold/Winter

At Home           Hot/Summer

At Observatory    Windy

On Road/Event     Calm

Company           Insects

Low Humidity      Clouds/Haze

High Humidity     Ground Fog

## Environmental Notes:

**Eyepiece Sketch**

## Personal Rating:

**1    2    3    4    5    6    7    8    9    10**

# Record
## of observation

**Date:**

**Time:**

| LOCATION | .................................... |
|---|---|
| CONSTELLATION | .................................... |
| OPTICS | .................................... |

| OBJECT | .................................... |
|---|---|
| FIRST SIGHTING | Y ☐ N ☐ |
| MAGNITUDE | .................................... |

## Observational Notes:

Seeing       Darkness

Transparency    Ambient Light

Moon       Cold/Winter

At Home    Hot/Summer

At Observatory   Windy

On Road/Event  Calm

Company    Insects

Low Humidity   Clouds/Haze

High Humidity  Ground Fog

## Environmental Notes:

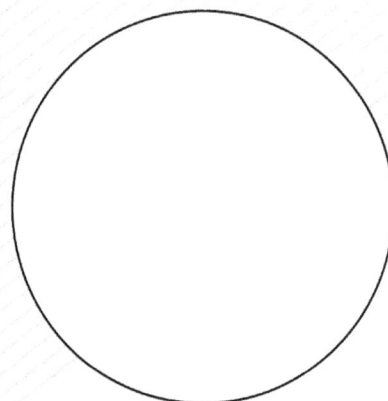

Eyepiece Sketch

## Personal Rating:

1    2    3    4    5    6    7    8    9    10

# Record
## of observation

**Date:**

**Time:**

| LOCATION | ............................... |
| CONSTELLATION | ............................... |
| OPTICS | ............................... |

| OBJECT | ............................... |
| FIRST SIGHTING | Y ☐ N ☐ |
| MAGNITUDE | ............................... |

| Seeing | Darkness |
|---|---|
| Transparency | Ambient Light |
| Moon | Cold/Winter |
| At Home | Hot/Summer |
| At Observatory | Windy |
| On Road/Event | Calm |
| Company | Insects |
| Low Humidity | Clouds/Haze |
| High Humidity | Ground Fog |

## Observational Notes:

## Environmental Notes:

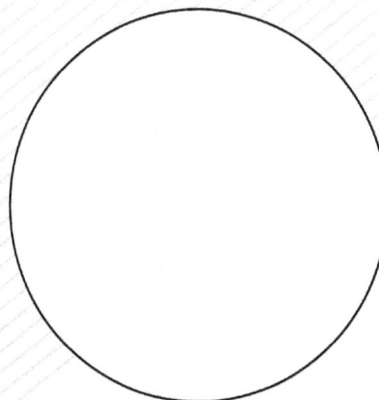

Eyepiece Sketch

## Personal Rating:

1    2    3    4    5    6    7    8    9    10

Josh Urban

# Record
## of observation

**Date:**

**Time:**

| LOCATION | ............................... |
| CONSTELLATION | ............................... |
| OPTICS | ............................... |

| OBJECT | ............................... |
| FIRST SIGHTING | Y ☐ N ☐ |
| MAGNITUDE | ............................... |

## Observational Notes:

| | |
|---|---|
| Seeing | Darkness |
| Transparency | Ambient Light |
| Moon | Cold/Winter |
| At Home | Hot/Summer |
| At Observatory | Windy |
| On Road/Event | Calm |
| Company | Insects |
| Low Humidity | Clouds/Haze |
| High Humidity | Ground Fog |

## Environmental Notes:

Eyepiece Sketch

**Personal Rating:**

1    2    3    4    5    6    7    8    9    10

# Record
## of observation

**Date:**

**Time:**

| LOCATION | ............................................... |
| CONSTELLATION | ............................................... |
| OPTICS | ............................................... |

| OBJECT | ............................................... |
| FIRST SIGHTING | Y ☐ N ☐ |
| MAGNITUDE | ............................................... |

Seeing          Darkness

Transparency    Ambient Light

Moon            Cold/Winter

At Home         Hot/Summer

At Observatory  Windy

On Road/Event   Calm

Company         Insects

Low Humidity    Clouds/Haze

High Humidity   Ground Fog

## Observational Notes:

## Environmental Notes:

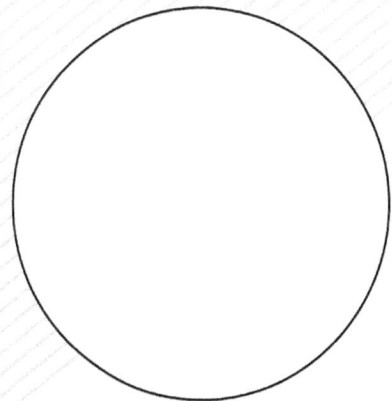

**Eyepiece Sketch**

## Personal Rating:

1      2      3      4      5      6      7      8      9      10

Josh Urban

# Record
## of observation

**Date:**

**Time:**

| | |
|---|---|
| **LOCATION** | ..................... |
| **CONSTELLATION** | ..................... |
| **OPTICS** | ..................... |

| | |
|---|---|
| **OBJECT** | ..................... |
| **FIRST SIGHTING** | Y ☐ N ☐ |
| **MAGNITUDE** | ..................... |

## Observational Notes:

| | |
|---|---|
| Seeing | Darkness |
| Transparency | Ambient Light |
| Moon | Cold/Winter |
| At Home | Hot/Summer |
| At Observatory | Windy |
| On Road/Event | Calm |
| Company | Insects |
| Low Humidity | Clouds/Haze |
| High Humidity | Ground Fog |

## Environmental Notes:

Eyepiece Sketch

## Personal Rating:

1   2   3   4   5   6   7   8   9   10

# Record
## of observation

**Date:**

**Time:**

| LOCATION | ......................................... |
| CONSTELLATION | ......................................... |
| OPTICS | ......................................... |

| OBJECT | ......................................... |
| FIRST SIGHTING | Y ☐ N ☐ |
| MAGNITUDE | ......................................... |

## Observational Notes:

Seeing          Darkness

Transparency    Ambient Light

Moon            Cold/Winter

At Home         Hot/Summer

At Observatory  Windy

On Road/Event   Calm

Company         Insects

Low Humidity    Clouds/Haze

High Humidity   Ground Fog

## Environmental Notes:

**Eyepiece Sketch**

## Personal Rating:

1    2    3    4    5    6    7    8    9    10

# Record
## of observation

**Date:**

**Time:**

| LOCATION | .................................. |
| CONSTELLATION | .................................. |
| OPTICS | .................................. |

| OBJECT | .................................. |
| FIRST SIGHTING | Y ☐ N ☐ |
| MAGNITUDE | .................................. |

## Observational Notes:

Seeing     Darkness

Transparency     Ambient Light

Moon     Cold/Winter

At Home     Hot/Summer

At Observatory     Windy

On Road/Event     Calm

Company     Insects

Low Humidity     Clouds/Haze

High Humidity     Ground Fog

## Environmental Notes:

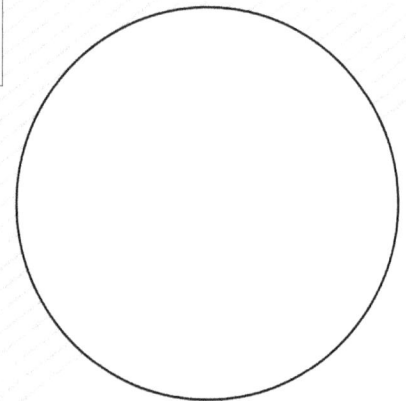

Eyepiece Sketch

## Personal Rating:

| 1 | 2 | 3 | 4 | 5 | 6 | 7 | 8 | 9 | 10 |

# Record
## of observation

**Date:**

**Time:**

| LOCATION | .............................................. |
| CONSTELLATION | .............................................. |
| OPTICS | .............................................. |

| OBJECT | .............................................. |
| FIRST SIGHTING | Y ☐ N ☐ |
| MAGNITUDE | .............................................. |

| Seeing | Darkness |
| Transparency | Ambient Light |
| Moon | Cold/Winter |
| At Home | Hot/Summer |
| At Observatory | Windy |
| On Road/Event | Calm |
| Company | Insects |
| Low Humidity | Clouds/Haze |
| High Humidity | Ground Fog |

## Observational Notes:

## Environmental Notes:

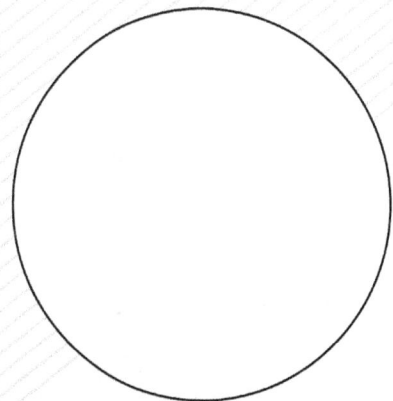

**Eyepiece Sketch**

## Personal Rating:

**1    2    3    4    5    6    7    8    9    10**

# Record
## of observation

**Date:**

**Time:**

| LOCATION | ............................................ |
| CONSTELLATION | ............................................ |
| OPTICS | ............................................ |

| OBJECT | ............................................ |
| FIRST SIGHTING | Y ☐ N ☐ |
| MAGNITUDE | ............................................ |

Seeing      Darkness

Transparency      Ambient Light

Moon      Cold/Winter

At Home      Hot/Summer

At Observatory      Windy

On Road/Event      Calm

Company      Insects

Low Humidity      Clouds/Haze

High Humidity      Ground Fog

## Observational Notes:

## Environmental Notes:

Eyepiece Sketch

**Personal Rating:**

1    2    3    4    5    6    7    8    9    10

# Record
## of observation

**Date:**

**Time:**

| LOCATION | .................................... |
| CONSTELLATION | .................................... |
| OPTICS | .................................... |

| OBJECT | .................................... |
| FIRST SIGHTING | Y ☐ N ☐ |
| MAGNITUDE | .................................... |

Seeing  Darkness

Transparency Ambient Light

Moon  Cold/Winter

At Home  Hot/Summer

At Observatory Windy

On Road/Event Calm

Company  Insects

Low Humidity Clouds/Haze

High Humidity Ground Fog

## Observational Notes:

## Environmental Notes:

**Eyepiece Sketch**

## Personal Rating:

1 2 3 4 5 6 7 8 9 10

# Record
## of observation

**Date:**

**Time:**

| | |
|---|---|
| **LOCATION** | .................................... |
| **CONSTELLATION** | .................................... |
| **OPTICS** | .................................... |

| | |
|---|---|
| **OBJECT** | .................................... |
| **FIRST SIGHTING** | Y ☐ N ☐ .................................... |
| **MAGNITUDE** | .................................... |

## Observational Notes:

| Seeing | Darkness |
|---|---|
| Transparency | Ambient Light |
| Moon | Cold/Winter |
| At Home | Hot/Summer |
| At Observatory | Windy |
| On Road/Event | Calm |
| Company | Insects |
| Low Humidity | Clouds/Haze |
| High Humidity | Ground Fog |

**Eyepiece Sketch**

## Environmental Notes:

## Personal Rating:

**1    2    3    4    5    6    7    8    9    10**

# Record
## of observation

**Date:**

**Time:**

| LOCATION | .................................................. |
| CONSTELLATION | .................................................. |
| OPTICS | .................................................. |

| OBJECT | .................................................. |
| FIRST SIGHTING | Y ☐ N ☐ |
| MAGNITUDE | .................................................. |

| Seeing | Darkness |
| Transparency | Ambient Light |
| Moon | Cold/Winter |
| At Home | Hot/Summer |
| At Observatory | Windy |
| On Road/Event | Calm |
| Company | Insects |
| Low Humidity | Clouds/Haze |
| High Humidity | Ground Fog |

## Observational Notes:

## Environmental Notes:

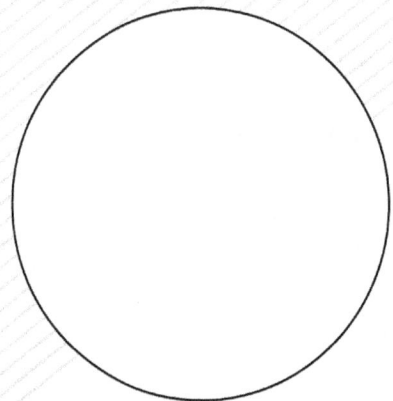

**Eyepiece Sketch**

## Personal Rating:

1    2    3    4    5    6    7    8    9    10

# Record
## of observation

**Date:**

**Time:**

| | |
|---|---|
| **LOCATION** | ................................................ |
| **CONSTELLATION** | ................................................ |
| **OPTICS** | ................................................ |

| | |
|---|---|
| **OBJECT** | ................................................ |
| **FIRST SIGHTING** | Y ☐ N ☐ |
| **MAGNITUDE** | ................................................ |

Seeing      Darkness

Transparency      Ambient Light

Moon      Cold/Winter

At Home      Hot/Summer

At Observatory      Windy

On Road/Event      Calm

Company      Insects

Low Humidity      Clouds/Haze

High Humidity      Ground Fog

## Observational Notes:

## Environmental Notes:

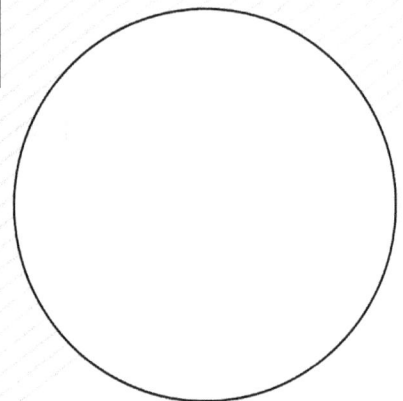

**Eyepiece Sketch**

**Personal Rating:**

| 1 | 2 | 3 | 4 | 5 | 6 | 7 | 8 | 9 | 10 |
|---|---|---|---|---|---|---|---|---|----|

# Record
## of observation

**Date:**

**Time:**

| LOCATION | .................................................. |
| CONSTELLATION | .................................................. |
| OPTICS | .................................................. |

| OBJECT | .................................................. |
| FIRST SIGHTING | Y ☐ N ☐ |
| MAGNITUDE | .................................................. |

## Observational Notes:

Seeing          Darkness

Transparency    Ambient Light

Moon            Cold/Winter

At Home         Hot/Summer

At Observatory  Windy

On Road/Event   Calm

Company         Insects

Low Humidity    Clouds/Haze

High Humidity   Ground Fog

## Environmental Notes:

Eyepiece Sketch

## Personal Rating:

1    2    3    4    5    6    7    8    9    10

# Record
## of observation

**Date:**

**Time:**

| LOCATION | .............................. | OBJECT | .............................. |
| CONSTELLATION | .............................. | FIRST SIGHTING | Y ☐ N ☐ |
| OPTICS | .............................. | MAGNITUDE | .............................. |

Seeing     Darkness

## Observational Notes:

Transparency     Ambient Light

Moon     Cold/Winter

At Home     Hot/Summer

At Observatory     Windy

On Road/Event     Calm

Company     Insects

Low Humidity     Clouds/Haze

High Humidity     Ground Fog

## Environmental Notes:

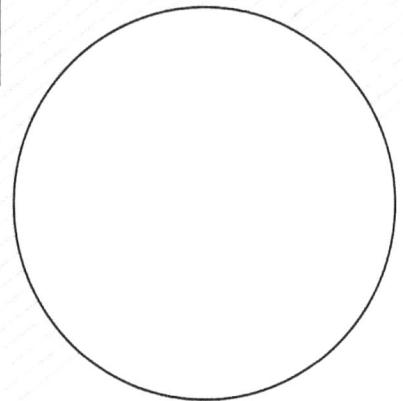

**Eyepiece Sketch**

## Personal Rating:

1    2    3    4    5    6    7    8    9    10

# Record
## of observation

**Date:**

**Time:**

| LOCATION | .................................................. |
| CONSTELLATION | .................................................. |
| OPTICS | .................................................. |

| OBJECT | .................................................. |
| FIRST SIGHTING | Y ☐ N ☐ .................................. |
| MAGNITUDE | .................................................. |

| Seeing | Darkness |
| Transparency | Ambient Light |
| Moon | Cold/Winter |
| At Home | Hot/Summer |
| At Observatory | Windy |
| On Road/Event | Calm |
| Company | Insects |
| Low Humidity | Clouds/Haze |
| High Humidity | Ground Fog |

## Observational Notes:

## Environmental Notes:

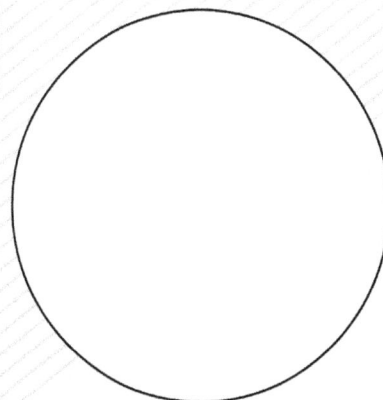

**Eyepiece Sketch**

## Personal Rating:

| 1 | 2 | 3 | 4 | 5 | 6 | 7 | 8 | 9 | 10 |

Josh Urban

# Record
## of observation

**Date:**

**Time:**

| LOCATION | ............................... |
| CONSTELLATION | ............................... |
| OPTICS | ............................... |

| OBJECT | ............................... |
| FIRST SIGHTING | Y ☐ N ☐ |
| MAGNITUDE | ............................... |

## Observational Notes:

| Seeing | Darkness |
| Transparency | Ambient Light |
| Moon | Cold/Winter |
| At Home | Hot/Summer |
| At Observatory | Windy |
| On Road/Event | Calm |
| Company | Insects |
| Low Humidity | Clouds/Haze |
| High Humidity | Ground Fog |

## Environmental Notes:

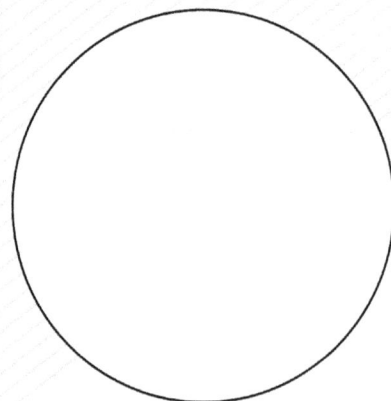

**Eyepiece Sketch**

## Personal Rating:

1    2    3    4    5    6    7    8    9    10

# Record
## of observation

**Date:**

**Time:**

| LOCATION | ............................................... |
| CONSTELLATION | ............................................... |
| OPTICS | ............................................... |

| OBJECT | ............................................... |
| FIRST SIGHTING | Y ☐ N ☐ |
| MAGNITUDE | ............................................... |

## Observational Notes:

Seeing          Darkness

Transparency    Ambient Light

Moon            Cold/Winter

At Home         Hot/Summer

At Observatory  Windy

On Road/Event   Calm

Company         Insects

Low Humidity    Clouds/Haze

High Humidity   Ground Fog

## Environmental Notes:

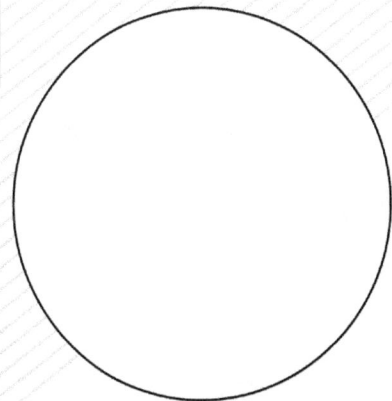

**Eyepiece Sketch**

## Personal Rating:

1    2    3    4    5    6    7    8    9    10

Josh Urban

# Record
## of observation

**Date:**

**Time:**

| LOCATION | ............................... | OBJECT | ............................... |
| CONSTELLATION | ............................... | FIRST SIGHTING | Y ☐ N ☐ |
| OPTICS | ............................... | MAGNITUDE | ............................... |

## Observational Notes:

Seeing | Darkness

Transparency | Ambient Light

Moon | Cold/Winter

At Home | Hot/Summer

At Observatory | Windy

On Road/Event | Calm

Company | Insects

Low Humidity | Clouds/Haze

High Humidity | Ground Fog

## Environmental Notes:

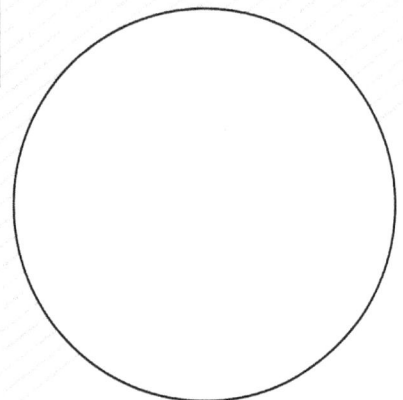

Eyepiece Sketch

**Personal Rating:**

1    2    3    4    5    6    7    8    9    10

# Record
## of observation

**Date:**

**Time:**

| LOCATION | .................................................... |
| CONSTELLATION | .................................................... |
| OPTICS | .................................................... |

| OBJECT | .................................................... |
| FIRST SIGHTING | Y ☐ N ☐ |
| MAGNITUDE | .................................................... |

Seeing · Darkness

Transparency · Ambient Light

Moon · Cold/Winter

At Home · Hot/Summer

At Observatory · Windy

On Road/Event · Calm

Company · Insects

Low Humidity · Clouds/Haze

High Humidity · Ground Fog

## Observational Notes:

## Environmental Notes:

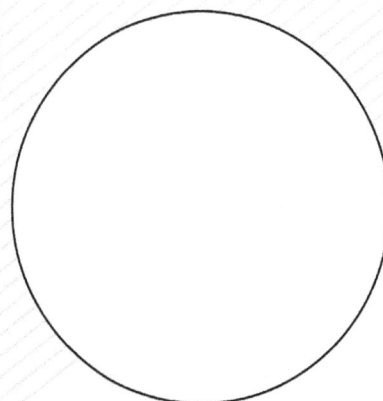

**Eyepiece Sketch**

## Personal Rating:

1    2    3    4    5    6    7    8    9    10

# Record
## of observation

**Date:**

**Time:**

| LOCATION | .................................................... |
| CONSTELLATION | .................................................... |
| OPTICS | .................................................... |

| OBJECT | .................................................... |
| FIRST SIGHTING | Y ☐ N ☐ .................................................... |
| MAGNITUDE | .................................................... |

Seeing            Darkness

Transparency      Ambient Light

Moon              Cold/Winter

At Home           Hot/Summer

At Observatory    Windy

On Road/Event     Calm

Company           Insects

Low Humidity      Clouds/Haze

High Humidity     Ground Fog

## Observational Notes:

## Environmental Notes:

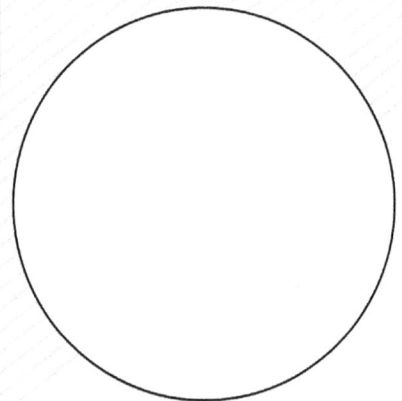

Eyepiece Sketch

## Personal Rating:

**1     2     3     4     5     6     7     8     9     10**

# Record
## of observation

**Date:**

**Time:**

| LOCATION | |
|---|---|
| **CONSTELLATION** | |
| **OPTICS** | |

| OBJECT | |
|---|---|
| **FIRST SIGHTING** | Y ☐ N ☐ |
| **MAGNITUDE** | |

Seeing          Darkness

Transparency    Ambient Light

Moon            Cold/Winter

At Home         Hot/Summer

At Observatory  Windy

On Road/Event   Calm

Company         Insects

Low Humidity    Clouds/Haze

High Humidity   Ground Fog

## Observational Notes:

## Environmental Notes:

Eyepiece Sketch

## Personal Rating:

1    2    3    4    5    6    7    8    9    10

Josh Urban

# Record
## of observation

**Date:**

**Time:**

| | | | |
|---|---|---|---|
| **LOCATION** | .................................. | **OBJECT** | .................................. |
| **CONSTELLATION** | .................................. | **FIRST SIGHTING** | Y ☐ N ☐ |
| **OPTICS** | .................................. | **MAGNITUDE** | .................................. |

## Observational Notes:

| | |
|---|---|
| Seeing | Darkness |
| Transparency | Ambient Light |
| Moon | Cold/Winter |
| At Home | Hot/Summer |
| At Observatory | Windy |
| On Road/Event | Calm |
| Company | Insects |
| Low Humidity | Clouds/Haze |
| High Humidity | Ground Fog |

## Environmental Notes:

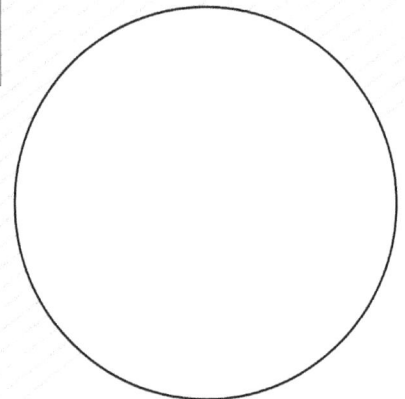

**Eyepiece Sketch**

## Personal Rating:

1    2    3    4    5    6    7    8    9    10

# Record
## of observation

**Date:**

**Time:**

| LOCATION | ............................................ |
| CONSTELLATION | ............................................ |
| OPTICS | ............................................ |

| OBJECT | ............................................ |
| FIRST SIGHTING | Y ☐ N ☐ |
| MAGNITUDE | ............................................ |

| Seeing | Darkness |
| Transparency | Ambient Light |
| Moon | Cold/Winter |
| At Home | Hot/Summer |
| At Observatory | Windy |
| On Road/Event | Calm |
| Company | Insects |
| Low Humidity | Clouds/Haze |
| High Humidity | Ground Fog |

## Observational Notes:

## Environmental Notes:

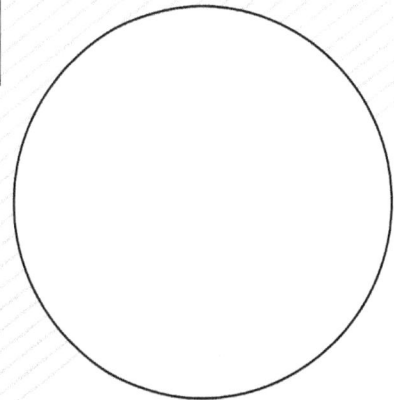

**Eyepiece Sketch**

## Personal Rating:

**1    2    3    4    5    6    7    8    9    10**

Josh Urban

# Record
## of observation

**Date:**

**Time:**

| LOCATION | |
| CONSTELLATION | |
| OPTICS | |

| OBJECT | |
| FIRST SIGHTING | Y ☐ N ☐ |
| MAGNITUDE | |

Seeing     Darkness

Transparency     Ambient Light

Moon     Cold/Winter

At Home     Hot/Summer

At Observatory     Windy

On Road/Event     Calm

Company     Insects

Low Humidity     Clouds/Haze

High Humidity     Ground Fog

## Observational Notes:

## Environmental Notes:

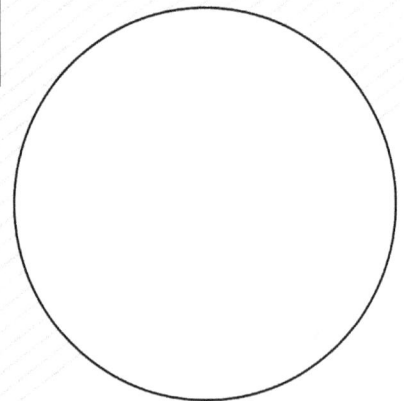

Eyepiece Sketch

## Personal Rating:

**1   2   3   4   5   6   7   8   9   10**

# Record
## of observation

**Date:**

**Time:**

| LOCATION | .................................................... |
| CONSTELLATION | .................................................... |
| OPTICS | .................................................... |

| OBJECT | .................................................... |
| FIRST SIGHTING | Y ☐ N ☐ |
| MAGNITUDE | .................................................... |

| Seeing | Darkness |
| Transparency | Ambient Light |
| Moon | Cold/Winter |
| At Home | Hot/Summer |
| At Observatory | Windy |
| On Road/Event | Calm |
| Company | Insects |
| Low Humidity | Clouds/Haze |
| High Humidity | Ground Fog |

## Observational Notes:

## Environmental Notes:

Eyepiece Sketch

## Personal Rating:

1    2    3    4    5    6    7    8    9    10

Josh Urban

# Record
## of observation

**Date:**

**Time:**

| LOCATION | ............................... |
| CONSTELLATION | ............................... |
| OPTICS | ............................... |

| OBJECT | ............................... |
| FIRST SIGHTING | Y ☐ N ☐ ............... |
| MAGNITUDE | ............................... |

| | |
|---|---|
| Seeing | Darkness |
| Transparency | Ambient Light |
| Moon | Cold/Winter |
| At Home | Hot/Summer |
| At Observatory | Windy |
| On Road/Event | Calm |
| Company | Insects |
| Low Humidity | Clouds/Haze |
| High Humidity | Ground Fog |

## Observational Notes:

## Environmental Notes:

**Eyepiece Sketch**

## Personal Rating:

1   2   3   4   5   6   7   8   9   10

# Record
## of observation

**Date:**

**Time:**

| LOCATION | ............................................. |
|---|---|
| CONSTELLATION | ............................................. |
| OPTICS | ............................................. |

| OBJECT | ............................................. |
|---|---|
| FIRST SIGHTING | Y ☐ N ☐ ............... |
| MAGNITUDE | ............................................. |

## Observational Notes:

Seeing      Darkness

Transparency      Ambient Light

Moon      Cold/Winter

At Home      Hot/Summer

At Observatory      Windy

On Road/Event      Calm

Company      Insects

Low Humidity      Clouds/Haze

High Humidity      Ground Fog

## Environmental Notes:

**Eyepiece Sketch**

## Personal Rating:

1    2    3    4    5    6    7    8    9    10

# Record
## of observation

**Date:**

**Time:**

| LOCATION | .................................................... |
| CONSTELLATION | .................................................... |
| OPTICS | .................................................... |

| OBJECT | .................................................... |
| FIRST SIGHTING | Y ☐ N ☐ |
| MAGNITUDE | .................................................... |

Seeing     Darkness

## Observational Notes:

Transparency     Ambient Light

Moon     Cold/Winter

At Home     Hot/Summer

At Observatory     Windy

On Road/Event     Calm

Company     Insects

Low Humidity     Clouds/Haze

High Humidity     Ground Fog

## Environmental Notes:

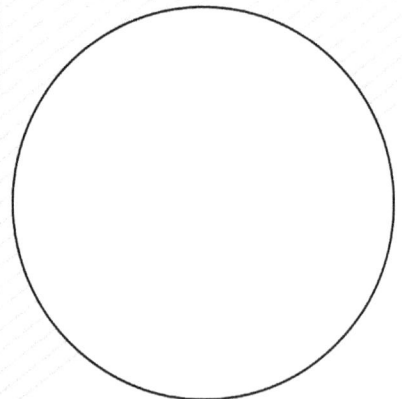

Eyepiece Sketch

## Personal Rating:

1    2    3    4    5    6    7    8    9    10

# Record
## of observation

**Date:**

**Time:**

| LOCATION | |
|---|---|
| CONSTELLATION | |
| OPTICS | |

| OBJECT | |
|---|---|
| FIRST SIGHTING | Y ☐ N ☐ |
| MAGNITUDE | |

Seeing      Darkness

Transparency      Ambient Light

Moon      Cold/Winter

At Home      Hot/Summer

At Observatory      Windy

On Road/Event      Calm

Company      Insects

Low Humidity      Clouds/Haze

High Humidity      Ground Fog

## Observational Notes:

## Environmental Notes:

**Eyepiece Sketch**

## Personal Rating:

1    2    3    4    5    6    7    8    9    10

# Record
## of observation

**Date:**

**Time:**

| LOCATION | ............................................. |
| CONSTELLATION | ............................................. |
| OPTICS | ............................................. |

| OBJECT | ............................................. |
| FIRST SIGHTING | Y ☐ N ☐ ............................. |
| MAGNITUDE | ............................................. |

## Observational Notes:

Seeing          Darkness

Transparency    Ambient Light

Moon            Cold/Winter

At Home         Hot/Summer

At Observatory  Windy

On Road/Event   Calm

Company         Insects

Low Humidity    Clouds/Haze

High Humidity   Ground Fog

## Environmental Notes:

**Eyepiece Sketch**

## Personal Rating:

1    2    3    4    5    6    7    8    9    10

# Record
## of observation

**Date:**

**Time:**

| LOCATION | ............................................. |
| CONSTELLATION | ............................................. |
| OPTICS | ............................................. |

| OBJECT | ............................................. |
| FIRST SIGHTING | Y ☐ N ☐ |
| MAGNITUDE | ............................................. |

## Observational Notes:

Seeing            Darkness

Transparency      Ambient Light

Moon              Cold/Winter

At Home           Hot/Summer

At Observatory    Windy

On Road/Event     Calm

Company           Insects

Low Humidity      Clouds/Haze

High Humidity     Ground Fog

## Environmental Notes:

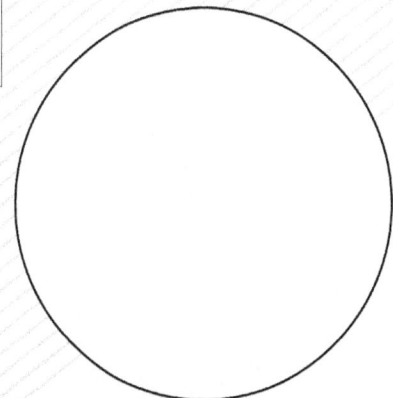

**Eyepiece Sketch**

## Personal Rating:

1    2    3    4    5    6    7    8    9    10

# Record
## of observation

**Date:**

**Time:**

| **LOCATION** | ............................................ |
| **CONSTELLATION** | ............................................ |
| **OPTICS** | ............................................ |

| **OBJECT** | ............................................ |
| **FIRST SIGHTING** | Y ☐ N ☐ |
| **MAGNITUDE** | ............................................ |

## Observational Notes:

| Seeing | Darkness |
| Transparency | Ambient Light |
| Moon | Cold/Winter |
| At Home | Hot/Summer |
| At Observatory | Windy |
| On Road/Event | Calm |
| Company | Insects |
| Low Humidity | Clouds/Haze |
| High Humidity | Ground Fog |

## Environmental Notes:

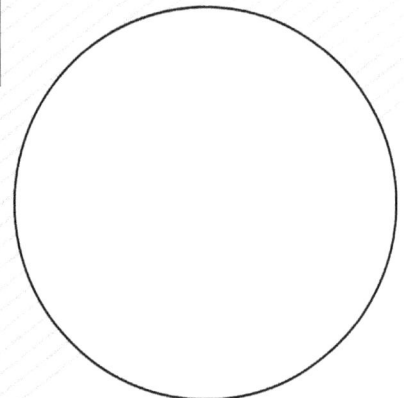

Eyepiece Sketch

**Personal Rating:**

| 1 | 2 | 3 | 4 | 5 | 6 | 7 | 8 | 9 | 10 |

# Record
## of observation

**Date:**

**Time:**

| LOCATION | ............................................... |
| CONSTELLATION | ............................................... |
| OPTICS | ............................................... |

| OBJECT | ............................................... |
| FIRST SIGHTING | Y ☐ N ☐ |
| MAGNITUDE | ............................................... |

## Observational Notes:

Seeing          Darkness

Transparency    Ambient Light

Moon            Cold/Winter

At Home         Hot/Summer

At Observatory  Windy

On Road/Event   Calm

Company         Insects

Low Humidity    Clouds/Haze

High Humidity   Ground Fog

## Environmental Notes:

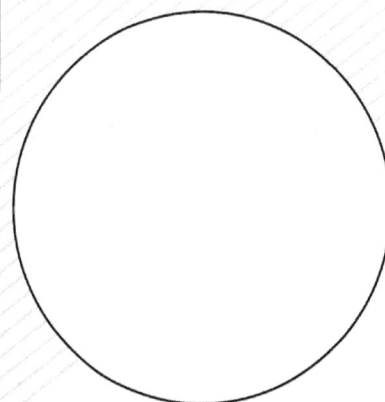

Eyepiece Sketch

## Personal Rating:

1    2    3    4    5    6    7    8    9    10

# Record
## of observation

**Date:**

**Time:**

| LOCATION | .................................... |
| CONSTELLATION | .................................... |
| OPTICS | .................................... |

| OBJECT | .................................... |
| FIRST SIGHTING | Y ☐ N ☐ |
| MAGNITUDE | .................................... |

Seeing     Darkness

Transparency     Ambient Light

Moon     Cold/Winter

At Home     Hot/Summer

At Observatory     Windy

On Road/Event     Calm

Company     Insects

Low Humidity     Clouds/Haze

High Humidity     Ground Fog

## Observational Notes:

## Environmental Notes:

**Eyepiece Sketch**

## Personal Rating:

1    2    3    4    5    6    7    8    9    10

# Record
## of observation

**Date:**

**Time:**

| LOCATION | .................................... |
| CONSTELLATION | .................................... |
| OPTICS | .................................... |

| OBJECT | .................................... |
| FIRST SIGHTING | Y ☐ N ☐ |
| MAGNITUDE | .................................... |

| Seeing | Darkness |
| Transparency | Ambient Light |
| Moon | Cold/Winter |
| At Home | Hot/Summer |
| At Observatory | Windy |
| On Road/Event | Calm |
| Company | Insects |
| Low Humidity | Clouds/Haze |
| High Humidity | Ground Fog |

## Observational Notes:

## Environmental Notes:

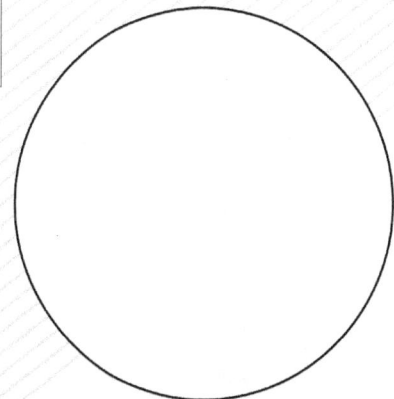

**Eyepiece Sketch**

## Personal Rating:

| 1 | 2 | 3 | 4 | 5 | 6 | 7 | 8 | 9 | 10 |

# Record
## of observation

**Date:**

**Time:**

| LOCATION | ........................................ |
| CONSTELLATION | ........................................ |
| OPTICS | ........................................ |

| OBJECT | ........................................ |
| FIRST SIGHTING | Y ☐ N ☐ |
| MAGNITUDE | ........................................ |

## Observational Notes:

Seeing          Darkness

Transparency    Ambient Light

Moon            Cold/Winter

At Home         Hot/Summer

At Observatory  Windy

On Road/Event   Calm

Company         Insects

Low Humidity    Clouds/Haze

High Humidity   Ground Fog

## Environmental Notes:

**Eyepiece Sketch**

## Personal Rating:

**1     2     3     4     5     6     7     8     9     10**

# Record
## of observation

**Date:**

**Time:**

| LOCATION | |
|---|---|
| CONSTELLATION | |
| OPTICS | |

| OBJECT | |
|---|---|
| FIRST SIGHTING | Y ☐ N ☐ |
| MAGNITUDE | |

Seeing          Darkness

Transparency          Ambient Light

## Observational Notes:

Moon          Cold/Winter

At Home          Hot/Summer

At Observatory          Windy

On Road/Event          Calm

Company          Insects

Low Humidity          Clouds/Haze

High Humidity          Ground Fog

## Environmental Notes:

Eyepiece Sketch

## Personal Rating:

1        2        3        4        5        6        7        8        9        10

# Record
## of observation

**Date:**

**Time:**

| LOCATION | .................................. |
| CONSTELLATION | .................................. |
| OPTICS | .................................. |

| OBJECT | .................................. |
| FIRST SIGHTING | Y ☐ N ☐ .................................. |
| MAGNITUDE | .................................. |

## Observational Notes:

Seeing     Darkness

Transparency     Ambient Light

Moon     Cold/Winter

At Home     Hot/Summer

At Observatory     Windy

On Road/Event     Calm

Company     Insects

Low Humidity     Clouds/Haze

High Humidity     Ground Fog

## Environmental Notes:

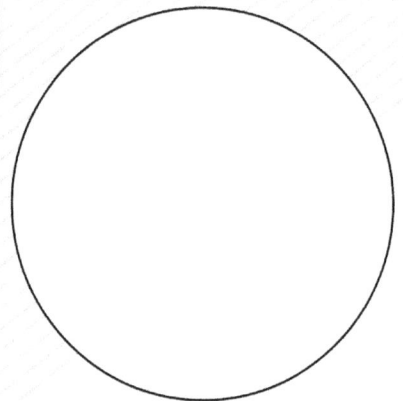

Eyepiece Sketch

## Personal Rating:

1    2    3    4    5    6    7    8    9    10

# Record
## of observation

**Date:**

**Time:**

| LOCATION | ...................... |
| CONSTELLATION | ...................... |
| OPTICS | ...................... |

| OBJECT | ...................... |
| FIRST SIGHTING | Y ☐ N ☐ |
| MAGNITUDE | ...................... |

Seeing          Darkness

Transparency    Ambient Light

Moon            Cold/Winter

At Home         Hot/Summer

At Observatory  Windy

On Road/Event   Calm

Company         Insects

Low Humidity    Clouds/Haze

High Humidity   Ground Fog

## Observational Notes:

## Environmental Notes:

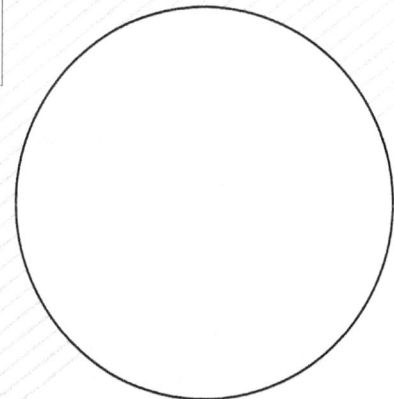

**Eyepiece Sketch**

## Personal Rating:

| 1 | 2 | 3 | 4 | 5 | 6 | 7 | 8 | 9 | 10 |

# Record
## of observation

**Date:**

**Time:**

| LOCATION | ............................... |
| CONSTELLATION | ............................... |
| OPTICS | ............................... |

| OBJECT | ............................... |
| FIRST SIGHTING | Y ☐ N ☐ ............................... |
| MAGNITUDE | ............................... |

Seeing     Darkness

Transparency     Ambient Light

Moon     Cold/Winter

At Home     Hot/Summer

At Observatory     Windy

On Road/Event     Calm

Company     Insects

Low Humidity     Clouds/Haze

High Humidity     Ground Fog

## Observational Notes:

## Environmental Notes:

**Eyepiece Sketch**

## Personal Rating:

1    2    3    4    5    6    7    8    9    10

# Record
## of observation

**Date:**

**Time:**

| LOCATION | ............................................... |
| CONSTELLATION | ............................................... |
| OPTICS | ............................................... |

| OBJECT | ............................................... |
| FIRST SIGHTING | Y ☐ N ☐ ............................... |
| MAGNITUDE | ............................................... |

| Seeing | Darkness |
| Transparency | Ambient Light |
| Moon | Cold/Winter |
| At Home | Hot/Summer |
| At Observatory | Windy |
| On Road/Event | Calm |
| Company | Insects |
| Low Humidity | Clouds/Haze |
| High Humidity | Ground Fog |

## Observational Notes:

## Environmental Notes:

Eyepiece Sketch

## Personal Rating:

1    2    3    4    5    6    7    8    9    10

# Record
## of observation

**Date:**

**Time:**

| LOCATION | ............................................... |
| CONSTELLATION | ............................................... |
| OPTICS | ............................................... |

| OBJECT | ............................................... |
| FIRST SIGHTING | Y ☐ N ☐ ............................... |
| MAGNITUDE | ............................................... |

## Observational Notes:

| Seeing | Darkness |
| Transparency | Ambient Light |
| Moon | Cold/Winter |
| At Home | Hot/Summer |
| At Observatory | Windy |
| On Road/Event | Calm |
| Company | Insects |
| Low Humidity | Clouds/Haze |
| High Humidity | Ground Fog |

## Environmental Notes:

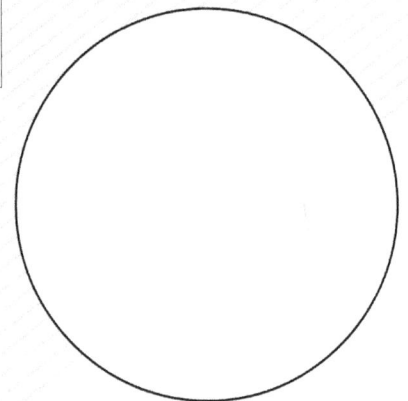

**Eyepiece Sketch**

## Personal Rating:

1    2    3    4    5    6    7    8    9    10

# Record
## of observation

**Date:**

**Time:**

| LOCATION | .................................................... |
| CONSTELLATION | .................................................... |
| OPTICS | .................................................... |

| OBJECT | .................................................... |
| FIRST SIGHTING | Y ☐ N ☐ |
| MAGNITUDE | .................................................... |

Seeing              Darkness

Transparency        Ambient Light

Moon                Cold/Winter

At Home             Hot/Summer

At Observatory      Windy

On Road/Event       Calm

Company             Insects

Low Humidity        Clouds/Haze

High Humidity       Ground Fog

## Observational Notes:

## Environmental Notes:

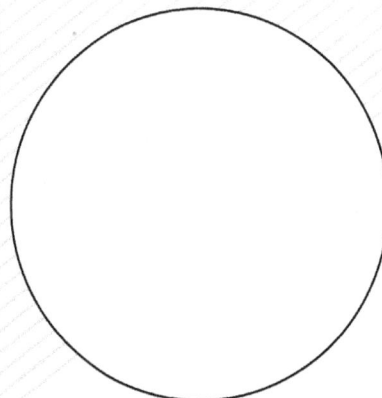

**Eyepiece Sketch**

## Personal Rating:

1     2     3     4     5     6     7     8     9     10

# Record
## of observation

**Date:**

**Time:**

| LOCATION | ........................................ | OBJECT | ........................................ |
| CONSTELLATION | ........................................ | FIRST SIGHTING | Y ☐ N ☐ ........................ |
| OPTICS | ........................................ | MAGNITUDE | ........................................ |

| Seeing | Darkness |
| Transparency | Ambient Light |
| Moon | Cold/Winter |
| At Home | Hot/Summer |
| At Observatory | Windy |
| On Road/Event | Calm |
| Company | Insects |
| Low Humidity | Clouds/Haze |
| High Humidity | Ground Fog |

## Observational Notes:

## Environmental Notes:

**Eyepiece Sketch**

## Personal Rating:

| 1 | 2 | 3 | 4 | 5 | 6 | 7 | 8 | 9 | 10 |

# Record
## of observation

**Date:**

**Time:**

| LOCATION | .................................... |
| CONSTELLATION | .................................... |
| OPTICS | .................................... |

| OBJECT | .................................... |
| FIRST SIGHTING | Y ☐ N ☐ .................... |
| MAGNITUDE | .................................... |

Seeing     Darkness

Transparency     Ambient Light

Moon     Cold/Winter

At Home     Hot/Summer

At Observatory     Windy

On Road/Event     Calm

Company     Insects

Low Humidity     Clouds/Haze

High Humidity     Ground Fog

## Observational Notes:

## Environmental Notes:

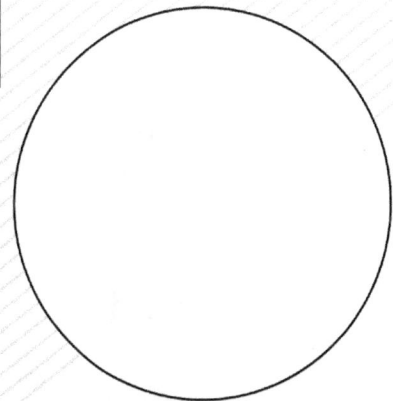

**Eyepiece Sketch**

## Personal Rating:

1    2    3    4    5    6    7    8    9    10

Josh Urban

# Record
## of observation

**Date:**

**Time:**

| LOCATION | ............................... |
| CONSTELLATION | ............................... |
| OPTICS | ............................... |

| OBJECT | ............................... |
| FIRST SIGHTING | Y ☐ N ☐ |
| MAGNITUDE | ............................... |

Seeing     Darkness

Transparency     Ambient Light

Moon     Cold/Winter

## Observational Notes:

At Home     Hot/Summer

At Observatory     Windy

On Road/Event     Calm

Company     Insects

Low Humidity     Clouds/Haze

High Humidity     Ground Fog

## Environmental Notes:

**Eyepiece Sketch**

## Personal Rating:

1   2   3   4   5   6   7   8   9   10

# Record
## of observation

**Date:**

**Time:**

| LOCATION | ............................................ |
| CONSTELLATION | ............................................ |
| OPTICS | ............................................ |

| OBJECT | ............................................ |
| FIRST SIGHTING | Y ☐ N ☐ |
| MAGNITUDE | ............................................ |

## Observational Notes:

| | |
|---|---|
| Seeing | Darkness |
| Transparency | Ambient Light |
| Moon | Cold/Winter |
| At Home | Hot/Summer |
| At Observatory | Windy |
| On Road/Event | Calm |
| Company | Insects |
| Low Humidity | Clouds/Haze |
| High Humidity | Ground Fog |

## Environmental Notes:

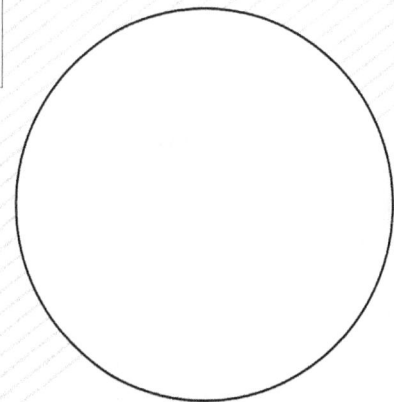

**Eyepiece Sketch**

## Personal Rating:

1    2    3    4    5    6    7    8    9    10

Josh Urban

# Record
## of observation

**Date:**

**Time:**

| LOCATION | .................................................... |
| CONSTELLATION | .................................................... |
| OPTICS | .................................................... |

| OBJECT | .................................................... |
| FIRST SIGHTING | Y ☐ N ☐ |
| MAGNITUDE | .................................................... |

Seeing     Darkness

## Observational Notes:

Transparency     Ambient Light

Moon     Cold/Winter

At Home     Hot/Summer

At Observatory     Windy

On Road/Event     Calm

Company     Insects

Low Humidity     Clouds/Haze

High Humidity     Ground Fog

## Environmental Notes:

Eyepiece Sketch

**Personal Rating:**

1    2    3    4    5    6    7    8    9    10

# Record
## of observation

**Date:**

**Time:**

| LOCATION | ............................................ |
| CONSTELLATION | ............................................ |
| OPTICS | ............................................ |

| OBJECT | ............................................ |
| FIRST SIGHTING | Y ☐ N ☐ |
| MAGNITUDE | ............................................ |

| Seeing | Darkness |
| Transparency | Ambient Light |
| Moon | Cold/Winter |
| At Home | Hot/Summer |
| At Observatory | Windy |
| On Road/Event | Calm |
| Company | Insects |
| Low Humidity | Clouds/Haze |
| High Humidity | Ground Fog |

## Observational Notes:

## Environmental Notes:

**Eyepiece Sketch**

## Personal Rating:

1    2    3    4    5    6    7    8    9    10

# Record
## of observation

**Date:**

**Time:**

| LOCATION | |
|---|---|
| CONSTELLATION | |
| OPTICS | |

| OBJECT | |
|---|---|
| FIRST SIGHTING | Y ☐ N ☐ |
| MAGNITUDE | |

Seeing          Darkness

Transparency    Ambient Light

Moon            Cold/Winter

At Home         Hot/Summer

At Observatory  Windy

On Road/Event   Calm

Company         Insects

Low Humidity    Clouds/Haze

High Humidity   Ground Fog

## Observational Notes:

## Environmental Notes:

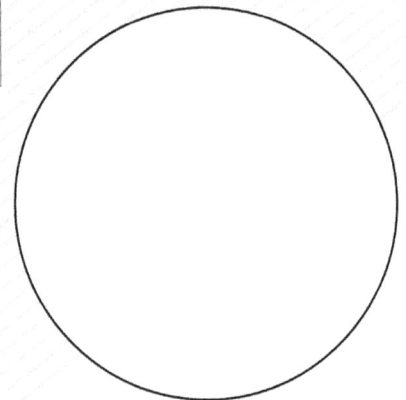

**Eyepiece Sketch**

## Personal Rating:

1    2    3    4    5    6    7    8    9    10

# Record
## of observation

**Date:**

**Time:**

| LOCATION | ............................................. |
| CONSTELLATION | ............................................. |
| OPTICS | ............................................. |

| OBJECT | ............................................. |
| FIRST SIGHTING | Y ☐ N ☐ |
| MAGNITUDE | ............................................. |

| Seeing | Darkness |
|---|---|
| Transparency | Ambient Light |
| Moon | Cold/Winter |
| At Home | Hot/Summer |
| At Observatory | Windy |
| On Road/Event | Calm |
| Company | Insects |
| Low Humidity | Clouds/Haze |
| High Humidity | Ground Fog |

## Observational Notes:

## Environmental Notes:

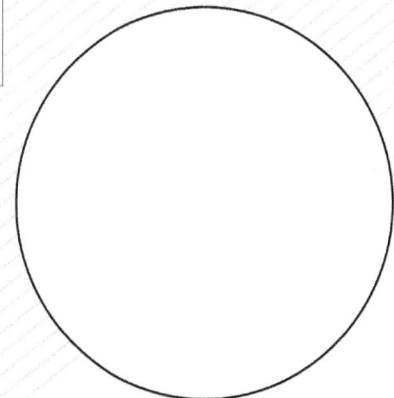

**Eyepiece Sketch**

## Personal Rating:

1    2    3    4    5    6    7    8    9    10

Josh Urban

# Record
## of observation

**Date:**

**Time:**

| LOCATION | ............................... |
| CONSTELLATION | ............................... |
| OPTICS | ............................... |

| OBJECT | ............................... |
| FIRST SIGHTING | Y ☐ N ☐ |
| MAGNITUDE | ............................... |

Seeing    Darkness

Transparency    Ambient Light

Moon    Cold/Winter

At Home    Hot/Summer

At Observatory    Windy

On Road/Event    Calm

Company    Insects

Low Humidity    Clouds/Haze

High Humidity    Ground Fog

## Observational Notes:

## Environmental Notes:

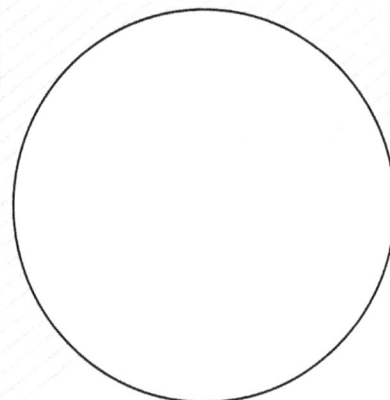

Eyepiece Sketch

**Personal Rating:**

1   2   3   4   5   6   7   8   9   10

# Record
## of observation

**Date:**

**Time:**

| LOCATION | .................................................. |
| CONSTELLATION | .................................................. |
| OPTICS | .................................................. |

| OBJECT | .................................................. |
| FIRST SIGHTING | Y ☐ N ☐ |
| MAGNITUDE | .................................................. |

## Observational Notes:

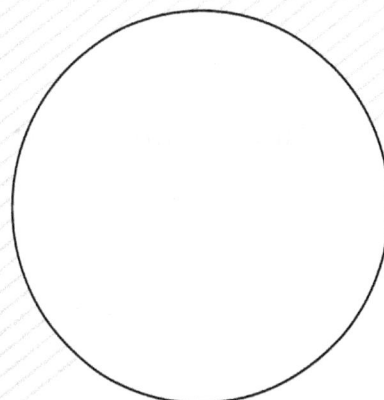

| Seeing | Darkness |
| Transparency | Ambient Light |
| Moon | Cold/Winter |
| At Home | Hot/Summer |
| At Observatory | Windy |
| On Road/Event | Calm |
| Company | Insects |
| Low Humidity | Clouds/Haze |
| High Humidity | Ground Fog |

## Environmental Notes:

**Eyepiece Sketch**

## Personal Rating:

1    2    3    4    5    6    7    8    9    10

# Record
## of observation

**Date:**

**Time:**

**LOCATION** ................................................

**CONSTELLATION** ................................................

**OPTICS** ................................................

**OBJECT** ................................................

**FIRST SIGHTING** Y ☐ N ☐ ................................................

**MAGNITUDE** ................................................

| | |
|---|---|
| Seeing | Darkness |
| Transparency | Ambient Light |
| Moon | Cold/Winter |
| At Home | Hot/Summer |
| At Observatory | Windy |
| On Road/Event | Calm |
| Company | Insects |
| Low Humidity | Clouds/Haze |
| High Humidity | Ground Fog |

## Observational Notes:

## Environmental Notes:

**Eyepiece Sketch**

## Personal Rating:

1　　2　　3　　4　　5　　6　　7　　8　　9　　10

# Record
## of observation

**Date:**

**Time:**

| LOCATION | ............................................. |
| CONSTELLATION | ............................................. |
| OPTICS | ............................................. |

| OBJECT | ............................................. |
| FIRST SIGHTING | Y ☐ N ☐ ............................................. |
| MAGNITUDE | ............................................. |

Seeing          Darkness

Transparency    Ambient Light

Moon            Cold/Winter

At Home         Hot/Summer

At Observatory  Windy

On Road/Event   Calm

Company         Insects

Low Humidity    Clouds/Haze

High Humidity   Ground Fog

## Observational Notes:

## Environmental Notes:

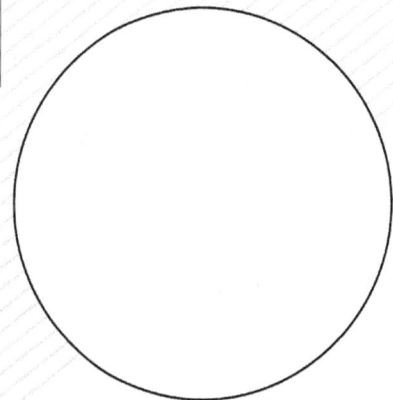

**Eyepiece Sketch**

## Personal Rating:

**1    2    3    4    5    6    7    8    9    10**

# Record
## of observation

**Date:**

**Time:**

| LOCATION | .................................. |
| CONSTELLATION | .................................. |
| OPTICS | .................................. |

| OBJECT | .................................. |
| FIRST SIGHTING | Y ☐ N ☐ |
| MAGNITUDE | .................................. |

Seeing            Darkness

Transparency      Ambient Light

Moon              Cold/Winter

At Home           Hot/Summer

At Observatory    Windy

On Road/Event     Calm

Company           Insects

Low Humidity      Clouds/Haze

High Humidity     Ground Fog

## Observational Notes:

## Environmental Notes:

**Eyepiece Sketch**

## Personal Rating:

1    2    3    4    5    6    7    8    9    10

# Record
## of observation

**Date:**

**Time:**

| LOCATION | |
| CONSTELLATION | |
| OPTICS | |

| OBJECT | |
| FIRST SIGHTING | Y ☐ N ☐ |
| MAGNITUDE | |

Seeing      Darkness

Transparency      Ambient Light

Moon      Cold/Winter

At Home      Hot/Summer

At Observatory      Windy

On Road/Event      Calm

Company      Insects

Low Humidity      Clouds/Haze

High Humidity      Ground Fog

## Observational Notes:

## Environmental Notes:

**Eyepiece Sketch**

## Personal Rating:

1    2    3    4    5    6    7    8    9    10

# Record
## of observation

**Date:**

**Time:**

| LOCATION | ............................ |
| CONSTELLATION | ............................ |
| OPTICS | ............................ |

| OBJECT | ............................ |
| FIRST SIGHTING | Y ☐ N ☐ |
| MAGNITUDE | ............................ |

| Seeing | Darkness |
| Transparency | Ambient Light |
| Moon | Cold/Winter |
| At Home | Hot/Summer |
| At Observatory | Windy |
| On Road/Event | Calm |
| Company | Insects |
| Low Humidity | Clouds/Haze |
| High Humidity | Ground Fog |

## Observational Notes:

## Environmental Notes:

**Eyepiece Sketch**

## Personal Rating:

1    2    3    4    5    6    7    8    9    10

# Record
## of observation

**Date:**

**Time:**

| LOCATION | .............................................. |
| CONSTELLATION | .............................................. |
| OPTICS | .............................................. |

| OBJECT | .............................................. |
| FIRST SIGHTING | Y ☐ N ☐ .............................................. |
| MAGNITUDE | .............................................. |

## Observational Notes:

| Seeing | Darkness |
| Transparency | Ambient Light |
| Moon | Cold/Winter |
| At Home | Hot/Summer |
| At Observatory | Windy |
| On Road/Event | Calm |
| Company | Insects |
| Low Humidity | Clouds/Haze |
| High Humidity | Ground Fog |

## Environmental Notes:

**Eyepiece Sketch**

## Personal Rating:

1    2    3    4    5    6    7    8    9    10

# Record
## of observation

**Date:**

**Time:**

| | |
|---|---|
| **LOCATION** | .................................... |
| **CONSTELLATION** | .................................... |
| **OPTICS** | .................................... |

| | |
|---|---|
| **OBJECT** | .................................... |
| **FIRST SIGHTING** | Y ☐ N ☐ |
| **MAGNITUDE** | .................................... |

| | |
|---|---|
| Seeing | Darkness |
| Transparency | Ambient Light |
| Moon | Cold/Winter |
| At Home | Hot/Summer |
| At Observatory | Windy |
| On Road/Event | Calm |
| Company | Insects |
| Low Humidity | Clouds/Haze |
| High Humidity | Ground Fog |

## Observational Notes:

## Environmental Notes:

**Eyepiece Sketch**

## Personal Rating:

**1    2    3    4    5    6    7    8    9    10**

# Record
## of observation

**Date:**

**Time:**

| LOCATION | .................................................... |
| CONSTELLATION | .................................................... |
| OPTICS | .................................................... |

| OBJECT | .................................................... |
| FIRST SIGHTING | Y ☐ N ☐ .................................... |
| MAGNITUDE | .................................................... |

Seeing          Darkness

Transparency    Ambient Light

Moon            Cold/Winter

At Home         Hot/Summer

At Observatory  Windy

On Road/Event   Calm

Company         Insects

Low Humidity    Clouds/Haze

High Humidity   Ground Fog

## Observational Notes:

## Environmental Notes:

**Eyepiece Sketch**

## Personal Rating:

1  2  3  4  5  6  7  8  9  10

# Record
## of observation

**Date:**

**Time:**

| | |
|---|---|
| **LOCATION** | .................................... |
| **CONSTELLATION** | .................................... |
| **OPTICS** | .................................... |

| | |
|---|---|
| **OBJECT** | .................................... |
| **FIRST SIGHTING** | Y ☐ N ☐ |
| **MAGNITUDE** | .................................... |

## Observational Notes:

Seeing          Darkness

Transparency    Ambient Light

Moon            Cold/Winter

At Home         Hot/Summer

At Observatory  Windy

On Road/Event   Calm

Company         Insects

Low Humidity    Clouds/Haze

High Humidity   Ground Fog

## Environmental Notes:

Eyepiece Sketch

## Personal Rating:

1    2    3    4    5    6    7    8    9    10

# Record
## of observation

**Date:**

**Time:**

| LOCATION | ............................................... |
| CONSTELLATION | ............................................... |
| OPTICS | ............................................... |

| OBJECT | ............................................... |
| FIRST SIGHTING | Y ☐ N ☐ ............................................... |
| MAGNITUDE | ............................................... |

## Observational Notes:

| Seeing | Darkness |
| Transparency | Ambient Light |
| Moon | Cold/Winter |
| At Home | Hot/Summer |
| At Observatory | Windy |
| On Road/Event | Calm |
| Company | Insects |
| Low Humidity | Clouds/Haze |
| High Humidity | Ground Fog |

## Environmental Notes:

**Eyepiece Sketch**

## Personal Rating:

1    2    3    4    5    6    7    8    9    10

Josh Urban

# Record
## of observation

**Date:**

**Time:**

| LOCATION | .................................... |
| CONSTELLATION | |
| OPTICS | |

| OBJECT | .................................... |
| FIRST SIGHTING | Y ☐ N ☐ |
| MAGNITUDE | |

Seeing　　　　　Darkness

Transparency　　Ambient Light

## Observational Notes:

Moon　　　　　Cold/Winter

At Home　　　　Hot/Summer

At Observatory　Windy

On Road/Event　Calm

Company　　　　Insects

Low Humidity　　Clouds/Haze

High Humidity　　Ground Fog

## Environmental Notes:

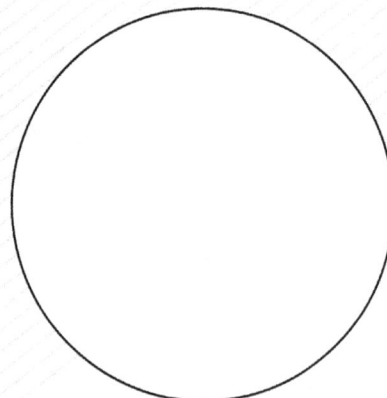

Eyepiece Sketch

## Personal Rating:

**1　2　3　4　5　6　7　8　9　10**

# Record
## of observation

**Date:**

**Time:**

| LOCATION | ............................................... |
| CONSTELLATION | ............................................... |
| OPTICS | ............................................... |

| OBJECT | ............................................... |
| FIRST SIGHTING | Y ☐ N ☐ |
| MAGNITUDE | ............................................... |

| Seeing | Darkness |
| Transparency | Ambient Light |
| Moon | Cold/Winter |
| At Home | Hot/Summer |
| At Observatory | Windy |
| On Road/Event | Calm |
| Company | Insects |
| Low Humidity | Clouds/Haze |
| High Humidity | Ground Fog |

## Observational Notes:

## Environmental Notes:

**Eyepiece Sketch**

## Personal Rating:

1    2    3    4    5    6    7    8    9    10

# Record
## of observation

**Date:**

**Time:**

| LOCATION | .................................... |
| CONSTELLATION | .................................... |
| OPTICS | .................................... |

| OBJECT | .................................... |
| FIRST SIGHTING | Y ☐ N ☐ |
| MAGNITUDE | .................................... |

| Seeing | Darkness |
| Transparency | Ambient Light |
| Moon | Cold/Winter |
| At Home | Hot/Summer |
| At Observatory | Windy |
| On Road/Event | Calm |
| Company | Insects |
| Low Humidity | Clouds/Haze |
| High Humidity | Ground Fog |

## Observational Notes:

## Environmental Notes:

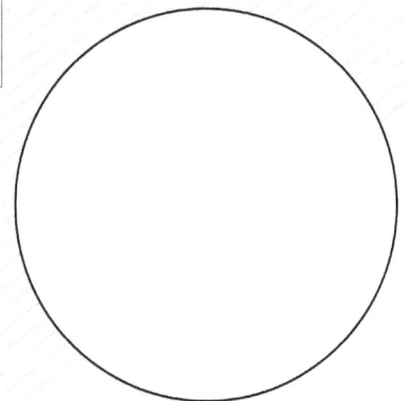

**Eyepiece Sketch**

## Personal Rating:

1    2    3    4    5    6    7    8    9    10

# Record
## of observation

**Date:**

**Time:**

| LOCATION | .................................................... |
| CONSTELLATION | .................................................... |
| OPTICS | .................................................... |

| OBJECT | .................................................... |
| FIRST SIGHTING | Y ☐ N ☐ |
| MAGNITUDE | .................................................... |

## Observational Notes:

| Seeing | Darkness |
| Transparency | Ambient Light |
| Moon | Cold/Winter |
| At Home | Hot/Summer |
| At Observatory | Windy |
| On Road/Event | Calm |
| Company | Insects |
| Low Humidity | Clouds/Haze |
| High Humidity | Ground Fog |

## Environmental Notes:

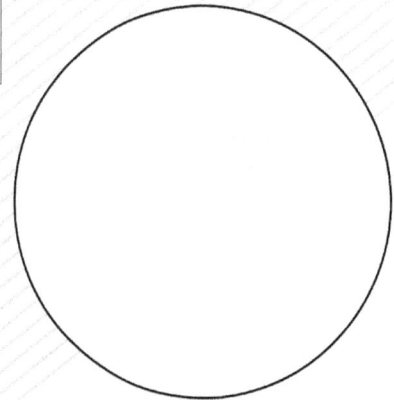

**Eyepiece Sketch**

## Personal Rating:

1    2    3    4    5    6    7    8    9    10

Josh Urban

# Record
## of observation

Date:

Time:

| LOCATION | ............... |
| CONSTELLATION | ............... |
| OPTICS | ............... |

OBJECT ..............

FIRST SIGHTING  Y ☐ N ☐

MAGNITUDE ..............

Seeing          Darkness

Transparency    Ambient Light

Moon            Cold/Winter

At Home         Hot/Summer

At Observatory  Windy

On Road/Event   Calm

Company         Insects

Low Humidity    Clouds/Haze

High Humidity   Ground Fog

## Observational Notes:

## Environmental Notes:

Eyepiece Sketch

## Personal Rating:

1    2    3    4    5    6    7    8    9    10

# Record
## of observation

**Date:**

**Time:**

| LOCATION | ............................................... |
| CONSTELLATION | ............................................... |
| OPTICS | ............................................... |

| OBJECT | ............................................... |
| FIRST SIGHTING | Y ☐ N ☐ |
| MAGNITUDE | ............................................... |

## Observational Notes:

Seeing          Darkness

Transparency    Ambient Light

Moon            Cold/Winter

At Home         Hot/Summer

At Observatory  Windy

On Road/Event   Calm

Company         Insects

Low Humidity    Clouds/Haze

High Humidity   Ground Fog

## Environmental Notes:

**Eyepiece Sketch**

## Personal Rating:

1    2    3    4    5    6    7    8    9    10

# Record
## of observation

**Date:**

**Time:**

| LOCATION | .................................... |
| CONSTELLATION | .................................... |
| OPTICS | .................................... |

| OBJECT | .................................... |
| FIRST SIGHTING | Y ☐ N ☐ |
| MAGNITUDE | .................................... |

| Seeing | Darkness |
| Transparency | Ambient Light |
| Moon | Cold/Winter |
| At Home | Hot/Summer |
| At Observatory | Windy |
| On Road/Event | Calm |
| Company | Insects |
| Low Humidity | Clouds/Haze |
| High Humidity | Ground Fog |

## Observational Notes:

## Environmental Notes:

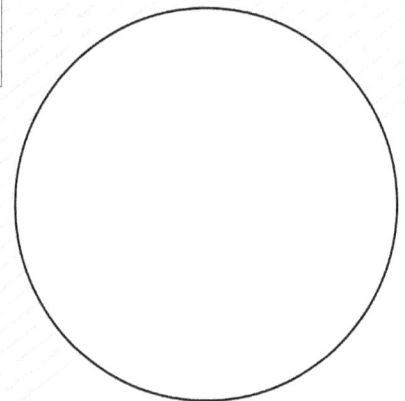

**Eyepiece Sketch**

## Personal Rating:

**1  2  3  4  5  6  7  8  9  10**

# Record
## of observation

**Date:**

**Time:**

| LOCATION | .................................... |
| CONSTELLATION | .................................... |
| OPTICS | .................................... |

| OBJECT | .................................... |
| FIRST SIGHTING | Y ☐ N ☐ .................................... |
| MAGNITUDE | .................................... |

| Seeing | Darkness |
| Transparency | Ambient Light |
| Moon | Cold/Winter |
| At Home | Hot/Summer |
| At Observatory | Windy |
| On Road/Event | Calm |
| Company | Insects |
| Low Humidity | Clouds/Haze |
| High Humidity | Ground Fog |

## Observational Notes:

## Environmental Notes:

**Eyepiece Sketch**

## Personal Rating:

1    2    3    4    5    6    7    8    9    10

# Record
## of observation

**Date:**

**Time:**

| LOCATION | ........................................ |
| CONSTELLATION | ........................................ |
| OPTICS | ........................................ |

| OBJECT | ........................................ |
| FIRST SIGHTING | Y ☐ N ☐ ........................... |
| MAGNITUDE | ........................................ |

## Observational Notes:

| Seeing | Darkness |
| Transparency | Ambient Light |
| Moon | Cold/Winter |
| At Home | Hot/Summer |
| At Observatory | Windy |
| On Road/Event | Calm |
| Company | Insects |
| Low Humidity | Clouds/Haze |
| High Humidity | Ground Fog |

## Environmental Notes:

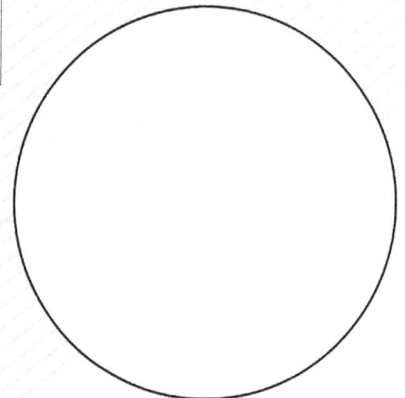

**Eyepiece Sketch**

## Personal Rating:

1    2    3    4    5    6    7    8    9    10

# Record
## of observation

**Date:**

**Time:**

| LOCATION | .................................................... |
| CONSTELLATION | .................................................... |
| OPTICS | .................................................... |

| OBJECT | .................................................... |
| FIRST SIGHTING | Y ☐ N ☐ |
| MAGNITUDE | .................................................... |

| Seeing | Darkness |
| Transparency | Ambient Light |
| Moon | Cold/Winter |
| At Home | Hot/Summer |
| At Observatory | Windy |
| On Road/Event | Calm |
| Company | Insects |
| Low Humidity | Clouds/Haze |
| High Humidity | Ground Fog |

## Observational Notes:

## Environmental Notes:

**Eyepiece Sketch**

## Personal Rating:

1    2    3    4    5    6    7    8    9    10

# Record
## of observation

**Date:**

**Time:**

| LOCATION | ............................... |
| CONSTELLATION | ............................... |
| OPTICS | ............................... |

| OBJECT | ............................... |
| FIRST SIGHTING | Y ☐ N ☐ ............................... |
| MAGNITUDE | ............................... |

Seeing          Darkness

## Observational Notes:

Transparency          Ambient Light

Moon          Cold/Winter

At Home          Hot/Summer

At Observatory          Windy

On Road/Event          Calm

Company          Insects

Low Humidity          Clouds/Haze

High Humidity          Ground Fog

## Environmental Notes:

Eyepiece Sketch

## Personal Rating:

1    2    3    4    5    6    7    8    9    10

# Chapter Three: Details

## Detailed Prompts and Abbreviations

**Size:** How "big" does the object look in the eyepiece? The Orion Nebula might occupy the entire field, while some tiny galaxies masquerade as fuzzy stars. Descriptions range from "really big!!" to formal arc measurements.

**Magnitude:** How bright is it? Catalog numbers can be misleading, as deep sky objects are *extended,* meaning the brightness of a single point is spread across the object's entire area. The larger the object, the more "stretched" the light becomes. Imagine: one drop of ink will appear vivid when concentrated in a blot, but a faint smudge if smeared across a sheet of paper. Add light pollution, lunar phase, and atmospheric conditions, and estimates get more difficult. I like to note how bright it looks to *me* for practical purposes, often using phrases instead of numbers. "Dim", "Barely Detected" "??", or, for the truly faint, "Averted Imagination."

**Texture/Structure:** Close study can tease out hidden detail. For example, the dim glow of a spiral galaxy might seem a ghost at first, but with time, do delicate mottling and hints of arms flicker into visibility? What do you notice?

**Context:** How does the "neighborhood" look? The dense star field surrounding M56 in Lyra reminds me of an alpine meadow in bloom. Scanning to the southeast, M30's lonely glow seems to bob in watery depths off the shore of Capricornus, the Sea Goat. While they're both globular clusters, the impression on the observer is radically different. Sometimes, there's a bonus prize nearby – the tiny galaxy NGC 6207 lurks like a celestial minnow around the M13 cluster in Hercules.

**Distance/Astrophysics:** Your notes don't have to cease when the sun rises. Try researching your targets after you log the initial observation. What's the distance to that galaxy? What's making that nebula glow? How old is that cluster?

**Conditions:** The weather, steadiness of the sky ("seeing") and transparency, lunar phase, light pollution, observing location, and terrestrial happenings are all noteworthy. Hey, if you see Bigfoot, write it down!

**People:** "How was your day?" Are you observing with friends? What's the human element?

**Josh's Excessively Poetic Suggestions:** If this object were a piece of music, what would it be? If it were a person, who does it recall? What nickname would it earn?

# Deep Sky Objects
Specific details to log with your observations might include:

*Galaxies:* Magnitude? Type? Apparent size in the eyepiece? Texture? Structure? Dust lanes? Context? Distance? Orientation (face on/edge on/tilted/etc?) *EG = Elliptical Galaxy SG = Spiral Galaxy IG = Irregular Galaxy (See "Abbreviations" for more.)*

*Diffuse Nebulae:* Magnitude? Size? Does a filter help? Structure/shape? Is it an emission, reflection, or a supernova remnant? How far across the sky can you trace it? Any hints of color? Interaction with other objects (e.g., dark or reflection nebulae found in the Trifid or Orion Nebulae complexes.) *EN = Emission Nebula RN = Reflection Nebula SNR = Supernova Remnant*

*Planetary Nebulae:* Magnitude? Color? Size? Structure? Central star? Best magnification? Context? Filter aids? *PN = Planetary Nebula*

*Dark Nebulae:* Size? Structure? Context and interaction with nearby objects and starfield? How far can it be traced? *DN = Dark Nebula*

*Open Clusters:* Magnitude? Size? Density? How many members? Shapes/pictures formed? Colors of members? Presence of doubles or variables? Best magnification? Context? *OC = Open Cluster*

*Globular Clusters:* Magnitude? Size? Density? Can it be resolved into individual stars, and if so, entirely to the core? What about star colors? Context? *GC = Globular Cluster*

*"The contemplation of celestial things will make a man both speak and think more sublimely and magnificently when he descends to human affairs."*
*– Cicero*

# Deep Sky Abbreviations

| Term | Useful Abbreviation* |
|------|---------------------|
| Galaxy | Gal |
| Elliptical Galaxy | EG |
| Spiral Galaxy | SG |
| Barred Spiral Galaxy | SBa, SBb, SBc (depending on structure) |
| Lenticular Galaxy | LG  (or S0) |
| Irregular Galaxy | IG |
| Emission Nebula | EN |
| Reflection Nebula | RN |
| Dark Nebula | DN |
| Supernova Remnant | SNR |
| Planetary Nebula | PN |
| Open Cluster | OC |
| Globular Cluster | GC |

*A balance has been struck between usefulness in the field and depth of detail.  Some subtypes have been omitted in the interest of usability.*

"The great spirals... Apparently lie outside our stellar system."
- Edwin Hubble

# The Constellations

| Constellation | Abbreviation | English Name |
| --- | --- | --- |
| Andromeda | And | The Chained Maiden |
| Antlia | Ant | The Air Pump |
| Apus | Aps | The Bird of Paradise |
| Aquarius | Aqr | The Water Bearer |
| Aquila | Aql | The Eagle |
| Ara | Ara | The Altar |
| Aries | Ari | The Ram |
| Auriga | Aur | The Charioteer |
| Boötes | Boo | The Herdsman |
| Caelum | Cae | The Engraving Tool |
| Camelopardalis | Cam | The Giraffe |
| Cancer | Cnc | The Crab |
| Canes Venatici | CVn | The Hunting Dogs |
| Canis Major | CMa | The Great Dog |
| Canis Minor | CMi | The Lesser Dog |
| Capricornus | Cap | The Sea Goat |
| Carina | Car | The Keel |
| Cassiopeia | Cas | The Queen |
| Centaurus | Cen | The Centaur |
| Cepheus | Cep | The King |
| Cetus | Cet | The Sea Monster |
| Chamaeleon | Cha | The Chameleon |
| Circinus | Cir | The Compass |
| Columba | Col | The Dove |
| Coma Berenices | Com | Berenice's Hair |
| Corona Australis | CrA | The Southern Crown |
| Cornoa Borealis | CrB | The Northern Crown |
| Corvus | Crv | The Crow |
| Crater | Crt | The Cup |

# The Constellations

| Constellation | Abbreviation | English Name |
|---|---|---|
| Crux | Cru | The Southern Cross |
| Cygnus | Cyg | The Swan |
| Delphinus | Del | The Dolphin |
| Dorado | Dor | The Dolphinfish |
| Draco | Dra | The Dragon |
| Equuleus | Eql | The Little Horse |
| Eridanus | Eri | The River |
| Fornax | For | The Furnace |
| Gemini | Gem | The Twins |
| Grus | Gru | The Crane |
| Hercules | Her | Hercules |
| Horologium | Hor | The Clock |
| Hydra | Hya | The Female Water Snake |
| Hydrus | Hyi | The Male Water Snake |
| Indus | Ind | The Indian |
| Lacerta | Lac | The Lizard |
| Leo | Leo | The Lion |
| Leo Minor | LMi | The Lesser Lion |
| Lepus | Lep | The Hare |
| Libra | Lib | The Scales |
| Lupus | Lup | The Wolf |
| Lynx | Lyn | The Lynx |
| Lyra | Lyr | The Lyre |
| Mensa | Men | The Table |
| Microscopium | Mic | The Microscope |
| Monoceros | Mon | The Unicorn |
| Musca | Mus | The Fly |
| Norma | Nor | The Carpenter's Square |
| Octans | Oct | The Octant |

# The Constellations

| Constellation | Abbreviation | English Name |
| --- | --- | --- |
| Ophiuchus | Oph | The Serpent Bearer |
| Orion | Ori | The Hunter |
| Pavo | Pav | The Peacock |
| Pegasus | Peg | The Winged Horse |
| Perseus | Per | The Hero |
| Phoenix | Phe | The Phoenix |
| Pictor | Pic | The Painter |
| Pisces | Psc | The Fishes |
| Piscis Austrinus | PsA | The Southern Fish |
| Puppis | Pup | The Stern |
| Pyxis | Pyx | The Compass |
| Reticulum | Ret | The Reticle |
| Sagitta | Sge | The Arrow |
| Sagittarius | Sgr | The Archer |
| Scorpius | Sco | The Scorpion |
| Sculptor | Scl | The Sculptor |
| Scutum | Sct | The Shield |
| Serpens | Ser | The Serpent |
| Sextans | Sex | The Sextant |
| Taurus | Tau | The Bull |
| Telescopium | Tel | The Telescope |
| Triangulum | Tri | The Triangle |
| Triangulum Australe | TrA | The Southern Triangle |
| Tucana | Tuc | The Toucan |
| Ursa Major | UMa | The Great Bear |
| Ursa Minor | UMi | The Lesser Bear |
| Vela | Vel | The Sails |
| Virgo | Vir | The Maiden |
| Volans | Vol | The Flying Fish |
| Vulpecula | Vul | The Fox |

# Chapter Four: Observing Projects

## Why

Sometimes "simply" getting lost in the stars is the best aim.  Other nights are ideal for a list of targets.  These lists can be helpful for special occasions, such as visiting a dark sky site.  If you're at a loss for what to observe, try the following projects on for size.

## The Messier List

The most popular of all observing projects is completing the Messier List.  Feel free to take years, or hit 'em all on one night at the next *Messier Marathon.*  Charles Messier (pronounced "messy a") first published his catalog of "false positives for comets" in 1771, adding to it as the years went by.  While hunting for these cosmic interlopers, he kept finding fuzzy blobs that *didn't* move, turning out to be nebulae, clusters, etc. Messier thought it helpful to make a list to avoid confusion, and today, these "inconveniences" rank as some of the finest sights for the backyard astronomer. You'll find a checklist in the Appendix.

## The Caldwell Objects

Generations after Messier weeded out comet impostors with his catalog, Sir Patrick Caldwell Moore made his own list of 109 essential objects.  First published in 1995, it contains sights not included by Messier, as well as many southern sky targets.  Some bring the heat:  Perseus' Double Cluster (Caldwell 14) or the Veil Nebula (Caldwell 33 & 34.).  Others are subtle, but scientifically interesting.  Caldwell 24 seems a dud at first.  Gazing through my 12.5" reflector, I was met with the bland haze of  NGC 1275. The tranquil sight hid the facts well:  It's a powerhouse of radio and magnetic emission, and violently spews out matter at relativistic speeds. The scale involved is dizzying: Caldwell 24 is about 250 *million* light years away!  We all know a sweet old uncle like that.  He sits so quietly in front of the TV that you'd never know he was a war hero.  Caldwell objects are especially suited to detailed notes and a further theoretical study in daylight hours.  (But then again, what isn't?)  Dig the checklist in the Appendix.

## Hipster Targets

Most telescopes end up parked on the top 40 hits of the night sky: The Andromeda Galaxy, M13, The Orion Nebula, etc. But what else is in the neighborhood? Try some B-sides and rarities. I call these "hipster targets." Scan a chart or your favorite app, and see if there's something you've missed before. Go for the obscure and overlooked! (And don't forget your vintage flannel while you're at it for extra credit.) I'll have to compile a list in the future. In the meantime, see if you can build your own.

## The Savor Project

I sometimes fall into "checklist madness." In the rush from splendor to splendor, things can blur and go unappreciated. If this sounds familiar, try an antidote: pick just *one* object to study for the evening. Tease out as much detail as you can. Make a sketch. If you have more than one set of optics, try different views, magnifications, and even the unaided eye. What's the smallest aperture you can detect it with? Does your finder scope show it? What's your *favorite* view? The Andromeda Galaxy is a fine example. A big scope can pick up extragalactic globular clusters and dust lanes. A small instrument provides a wider perspective, as companion galaxies swim into view. Binoculars reveal the stunning swath and extent of the arms. The naked eye glimpse of the "nearby" galaxy is thrilling in its own right, setting it in cosmic context. The timeless legends of heroes and vanity play out high overhead on quiet autumn evenings, as the crickets take their final encore.

## Keep Going

Once you've filled up this volume - don't stop! Create your own notebook, or get another copy of this one. The front page has a catalog number so you can start your own library of records. Imagine documenting a decade! Use it, smudge it, get it covered in dew and grass stains, and drench it in starlight.

## Don't Overdo it

Although taking notes *can* be helpful, don't get trapped. The priority is to enjoy the sky. I urge you to apply this book when helpful, ***but set it aside when it's not.*** Remember: ***look up!***

"*Man must rise above the Earth—to the top of the atmosphere and beyond—for only thus will he fully understand the world in which he lives.*"

*– Socrates*

# Chapter Five: Resources

## Further Reading

The following resources have aided me tremendously in my exploration of the heavens. You might find them to be of service, too.

## Charts

*Uranometria 2000.0* - Tirion/Rappaport/Remaklus
*Sky Atlas 2000.0* - Will Tirion
*Deep Map 600* - Will Tirion, available from Orion Telescopes

## Observing Guides

*Turn Left at Orion* - Guy Consolmagno
*The Messier Objects* - Stephen James O'Meara
*The Caldwell Objects* - Stephen James O'Meara
*Hidden Treasures* - Stephen James O'Meara
*The Night Sky Observer's Guide* - Kepple/Sanner
*Burnham's Celestial Handbook* - Robert Burnham Jr.

## Software

*Stellarium* (Free and open source!) www.stellarium.org

Don't forget the many useful apps for phones and tablets, both free and paid.

**Find what works for *you*.**

## Thank you and Goodnight

It's my hope that this guide has made you a better observer and you've seen the "impossible." Remember to be patient and enjoy the process. Learn a bit more of the sky each night, and don't forget to cruise the heavens with your binoculars. In fact, you might join the ranks of observers who use them exclusively (and to great effect.) Best wishes for many starry nights, and pleasant reading of your cosmic voyages when the rain is falling. Look, marvel, note, sketch, ponder, and study. Write to make blind men see. As you journey, let me know if you have any suggestions or tips to help future readers plumb the depths of the night.

Clear skies, and happy sailing!

- *Josh*

*"Astronomy compels the soul to look upwards and leads us from this world to another."*
*- Plato*

# Appendix: Checklists

## The Messier Objects

| Messier Number | Constellation | Name |
|---|---|---|
| 1 | Tau | Crab Nebula |
| 2 | Aqu | |
| 3 | CVn | |
| 4 | Sco | |
| 5 | Ser | |
| 6 | Sco | Butterfly Cluster |
| 7 | Sco | Ptolemy's Cluster |
| 8 | Sgr | Lagoon Nebula |
| 9 | Oph | |
| 10 | Oph | |
| 11 | Scu | Wild Duck Cluster |
| 12 | Oph | |
| 13 | Her | Hercules Cluster |
| 14 | Oph | |
| 15 | Peg | |
| 16 | Ser | Eagle Nebula |
| 17 | Sgr | Swan/Omega Nebula |
| 18 | Sgr | |
| 19 | Oph | |
| 20 | Sgr | Trifid Nebula |
| 21 | Sgr | |
| 22 | Sgr | |
| 23 | Sgr | |
| 24 | Sgr | Sagittarius Star Cloud |
| 25 | Sgr | |
| 26 | Scu | |
| 27 | Vul | Dumbbell Nebula |

# The Messier Objects

| Messier Number | Constellation/Target Type | Name |
|---|---|---|
| 28 | Sgr | |
| 29 | Cyg | |
| 30 | Cap | |
| 31 | And | Andromeda Galaxy |
| 32 | And | |
| 33 | Tri | Triangulum Galaxy |
| 34 | Per | |
| 35 | Gem | |
| 36 | Aur | |
| 37 | Aur | |
| 38 | Aur | |
| 39 | Cyg | |
| 40 | UMa | |
| 41 | CMa | |
| 42 | Ori | Orion Nebula |
| 43 | Ori | De Mairan's Nebula |
| 44 | Cnc | Beehive/Praesepe |
| 45 | Tau | Pleiades |
| 46 | Pup | |
| 47 | Pup | |
| 48 | Hya | |
| 49 | Vir | |
| 50 | Mon | |
| 51 | CVn | Whirlpool Galaxy |
| 52 | Cas | |
| 53 | Com | |
| 54 | Sgr | |
| 55 | Sgr | |

# The Messier Objects

| Messier Number | Constellation/Target Type | Name |
| --- | --- | --- |
| 56 | Lyr | |
| 57 | Lyr | Ring Nebula |
| 58 | Vir | |
| 59 | Vir | |
| 60 | Vir | |
| 61 | Vir | |
| 62 | Oph | |
| 63 | CVn | Sunflower Galaxy |
| 64 | Com | Black Eye Galaxy |
| 65 | Leo | |
| 66 | Leo | |
| 67 | Cnc | |
| 68 | Hya | |
| 69 | Sgr | |
| 70 | Sgr | |
| 71 | Sag | |
| 72 | Aqu | |
| 73 | Aqu | |
| 74 | Psc | Phantom Galaxy |
| 75 | Sgr | |
| 76 | Per | Little Dumbbell Nebula |
| 77 | Cet | |
| 78 | Ori | |
| 79 | Lep | |
| 80 | Sco | |
| 81 | UMa | Bode's Galaxy |
| 82 | UMa | Cigar Galaxy |

## The Messier Objects

| Messier Number | Constellation/Target Type | Name |
|---|---|---|
| 83 | Hya | Southern Pinwheel |
| 84 | Vir | |
| 85 | Com | |
| 86 | Vir | |
| 87 | Vir | Virgo A |
| 88 | Com | |
| 89 | Vir | |
| 90 | Vir | |
| 91 | Com | |
| 92 | Her | |
| 93 | Pup | |
| 94 | CVn | |
| 95 | Leo | |
| 96 | Leo | |
| 97 | UMa | Owl Nebula |
| 98 | Com | |
| 99 | Com | |
| 100 | Com | |
| 101 | UMa | Pinwheel Galaxy |
| 102 | Dra | Spindle Galaxy |
| 103 | Cas | |
| 104 | Vir | Sombrero Galaxy |
| 105 | Leo | |
| 106 | CVn | |
| 107 | Oph | |
| 108 | UMa | |
| 109 | UMa | |
| 110 | And | |

# The Caldwell Objects

| Caldwell Number | Catalog Number | Constellation |
|---|---|---|
| 1 | NGC 188 | Cep |
| 2 | NGC 40 | Cep |
| 3 | NGC 4236 | Dra |
| 4 | NGC 7023 | Cep |
| 5 | IC 342 | Cam |
| 6 | NGC 6543 | Dra |
| 7 | NGC 2403 | Cam |
| 8 | NGC 559 | Cas |
| 9 | Sh2-155 | Cep |
| 10 | NGC 663 | Cas |
| 11 | NGC 7635 | Cas |
| 12 | NGC 6946 | Cep |
| 13 | NGC 457 | Cas |
| 14 | NGC 869/884 | Per |
| 15 | NGC 6826 | Cyg |
| 16 | NGC 7243 | Lac |
| 17 | NGC 147 | Cas |
| 18 | NGC 185 | Cas |
| 19 | IC 5146 | Cyg |
| 20 | NGC 7000 | Cyg |
| 21 | NGC 4449 | CVn |
| 22 | NGC 7662 | And |
| 23 | NGC 891 | And |
| 24 | NGC 1275 | Per |
| 25 | NGC 2419 | Lyn |
| 26 | NGC 4244 | CVn |
| 27 | NGC 6888 | Cyg |
| 28 | NGC 752 | And |

# The Caldwell Objects

| Caldwell Number | Catalog Number | Constellation |
|---|---|---|
| 29 | NGC 5005 | CVn |
| 30 | NGC 7331 | Peg |
| 31 | IC 405 | Aur |
| 32 | NGC 4631 | CVn |
| 33 | NGC 6992/5 | Cyg |
| 34 | NGC 6960 | Cyg |
| 35 | NGC 4889 | Com |
| 36 | NGC 4559 | Com |
| 37 | NGC 6885 | Vul |
| 38 | NGC 4565 | Com |
| 39 | NGC 2392 | Gem |
| 40 | NGC 3626 | Leo |
| 41 | Mel 25 | Tau |
| 42 | NGC 7006 | Del |
| 43 | NGC 7814 | Peg |
| 44 | NGC 7479 | Peg |
| 45 | NGC 5248 | Boo |
| 46 | NGC 2261 | Mon |
| 47 | NGC 6934 | Del |
| 48 | NGC 2775 | Cnc |
| 49 | NGC 2237-8/46 | Mon |
| 50 | NGC 2244 | Mon |
| 51 | IC 1613 | Cet |
| 52 | NGC 4697 | Vir |
| 53 | NGC 3115 | Sex |
| 54 | NGC 2506 | Mon |
| 55 | NGC 7009 | Aqr |
| 56 | NGC 246 | Cet |

# The Caldwell Objects

| Caldwell Number | Catalog Number | Constellation |
|---|---|---|
| 57 | NGC 6822 | Sgr |
| 58 | NGC 2360 | CMa |
| 59 | NGC 3242 | Hya |
| 60 | NGC 4038 | Crv |
| 61 | NGC 4039 | Crv |
| 62 | NGC 247 | Cet |
| 63 | NGC 7293 | Aqr |
| 64 | NGC 2362 | CMa |
| 65 | NGC 253 | Scl |
| 66 | NGC 5694 | Hya |
| 67 | NGC 1097 | For |
| 68 | NGC 6729 | CrA |
| 69 | NGC 6302 | Sco |
| 70 | NGC 300 | Scl |
| 71 | NGC 2477 | Pup |
| 72 | NGC 55 | Scl |
| 73 | NGC 1851 | Col |
| 74 | NGC 3132 | Vel |
| 75 | NGC 6124 | Sco |
| 76 | NGC 6231 | Sco |
| 77 | NGC 5128 | Cen |
| 78 | NGC 6541 | CrA |
| 79 | NGC 3201 | Vel |
| 80 | NGC 5139 | Cen |
| 81 | NGC 6352 | Ara |
| 82 | NGC 6193 | Ara |
| 83 | NGC 4945 | Cen |
| 84 | NGC 5286 | Cen |

# The Caldwell Objects

| Caldwell Number | Catalog Number | Constellation |
|---|---|---|
| 85 | IC 2391 | Vel |
| 86 | NGC 6397 | Ara |
| 87 | NGC 1261 | Hor |
| 88 | NGC 5823 | Cir |
| 89 | NGC 6087 | Nor |
| 90 | NGC 2867 | Car |
| 91 | NGC 3532 | Car |
| 92 | NGC 3372 | Car |
| 93 | NGC 6752 | Pav |
| 94 | NGC 4755 | Cru |
| 95 | NGC 6025 | TrA |
| 96 | NGC 2516 | Car |
| 97 | NGC 3766 | Cen |
| 98 | NGC 4609 | Cru |
| 99 | Coalsack (Dark Nebula) | Cru |
| 100 | Cr 249 | Cen |
| 101 | NGC 6744 | Pav |
| 102 | IC 2602 | Car |
| 103 | NGC 2070 | Dor |
| 104 | NGC 362 | Tuc |
| 105 | NGC 4833 | Mus |
| 106 | NCG 104 | Tuc |
| 107 | NGC 6101 | Aps |
| 108 | NGC 4372 | Mus |
| 109 | NGC 3195 | Cha |

*"To know that we know what we know, and to know that we do not know what we do not know, that is true knowledge."*
*-Nicolaus Copernicus*

Josh Urban

# About the Author

**Josh Urban** is an astronomer, writer, and speaker. He's on a mission to share the thrill of the cosmos with his fellow Earthlings. With publications in *Sky and Telescope*, *Astronomy, Positive Impact*, and others, Urban is also the founder of *The Nighthawk Bulletin*, and *The Night Kitchen* podcast.

Outreach is a vital part of his work, especially to audiences who might not otherwise partake. Educational partners have included The Maryland School for the Blind, The National Air and Space Museum, The Maryland State Library for the Blind and Print Disabled, and many senior retirement communities. He lectures regularly on the topic. Showing people Saturn through a telescope remains a favorite activity.

He lives near Lynchburg, VA, USA with a collection of records and telescopes. Say hello at www.JoshUrban.com

www.ingramcontent.com/pod-product-compliance
Lightning Source LLC
Chambersburg PA
CBHW052113020426
42335CB00021B/2740